Civilian Conservation Corps

Recent Titles in
Bibliographies and Indexes in American History

The Kickapoo Indians, Their History and Culture: An Annotated Bibliography
Phillip M. White, compiler

The American Settlement Movement: A Bibliography
Domenica M. Barbuto, compiler

Reconstruction in the United States: An Annotated Bibliography
David A. Lincove, compiler and annotator

Books on Early American History and Culture, 1991-1995: An Annotated Bibliography
Raymond D. Irwin

Peyotism and the Native American Church: An Annotated Bibliography
Phillip M. White

A Comprehensive Catalogue of the Correspondence and Papers of James Monroe, Volume I
Daniel Preston

A Comprehensive Catalogue of the Correspondence and Papers of James Monroe, Volume II
Daniel Preston

Books on Early American History and Culture, 1986-1990: An Annotated Bibliography
Raymond D. Irwin

American Indian and African American People, Communities, and Interactions: An Annotated Bibliography
Lisa Bier

John Wesley Powell: An Annotated Bibliography
Marcia L. Thomas

Books on Early American History and Culture, 1971-1980: An Annotated Bibliography
Raymond D. Irwin

Books on Early American History and Culture, 1981-1985: An Annotated Bibliography
Raymond D. Irwin

Civilian Conservation Corps

A Selectively Annotated Bibliography

Larry N. Sypolt

Foreword by Harry D. Dallas

Bibliographies and Indexes in American History, Number 52

PRAEGER

Westport, Connecticut
London

Library of Congress Cataloging-in-Publication Data is available at www.loc.gov

British Library Cataloguing in Publication Data is available.

ISBN: 0-313-31313-X
ISSN: 0742-6828

First published in 2005

Praeger Publishers, 88 Post Road West, Westport, CT 06881
An imprint of Greenwood Publishing Group, Inc.
www.praeger.com

Printed in the United States of America

The paper used in this book complies with the Permanent Paper Standard issued by the National Information Standards Organization (Z39.48-1984).

10 9 8 7 6 5 4 3 2 1

To
All former enrollees,
military, and overhead personnel
for the lasting legacy they have left
for generations to enjoy.

And
To the memory of my brother
Lt.Col. Gary N. Sypolt

Contents

Foreword

The decade of the 1930s was a period of unrest and change for the United States. There were thousands of young men roaming the streets of the cities, "riding the rails" and living in hobo jungles. The drought in the mid-west caused dust storms and forced small towns and farming communities to become nearly bankrupt. Several hundred years of ravaging the nation's forests also contributed to dust storm and erosion problems. Factories and banks were closing.

President Franklin D. Roosevelt was elected in November 1932, and sworn into office on March 3, 1933. He was the father of the "Alphabet Agencies" such as the WPA, NYA, PWA, REA, TVA and CCC-the Civilian Conservation Corps. The most popular and successful of these programs was the CCC.

The CCC was created during the first hundred days of Roosevelt's administration. It was designed to take the unemployed youth off of the streets and provide worthwhile employment for them. They were organized into work companies and given work assignments in conservation camps. The camps were under the direction of the War Department. The enrollees were selected by the Labor Department, and agencies within the Department of the Interior and Department of Agriculture were in charge of the work assignments.

There were many different types of projects to which the camps were assigned. There were forestry, soil conservation, national parks, state parks and many other types of projects. Billions of trees were planted, miles of roads and trails were built, and many pavilions, picnic areas, lakes, dams and miles of fences constructed.

I am a former member of the CCC. I served in two camps in Illinois as an enrollee, assistant leader, leader and senior leader, between 1939 and 1941. I have participated in the greatest era of expansion in our state and national parks and forests.

This bibliography will assist students, scholars and historians and private individuals in researching the legacy of the CCC. The national focus allows the user to find references on local, state or federal levels. It also provides resources that tell where to find other CCC materials.

Harry D. Dallas
NACCCA Archivist

Preface

I became familiar with the Civilian Conservation Corps while doing an internship for a graduate degree in history at West Virginia University. The internship project was to write the history of the CCC camp at Fort Necessity, in Farmington, Pennsylvania. I was to undertake an oral history project where I interviewed former enrollees, officers and civilian staff from the CCC companies located at Fort Necessity, and, together with surviving camp newspapers, write the administrative history of the camp.

The internship was started in the mid 1980s, before the world wide web and computer technology as we know it today. I was fortunate to have lists of company enrollees and overhead personnel and their recent addresses, because of a reunion that was held a year earlier. However, I still had to correspond by mail or telephone with many of them, because they had not settled in the immediate vicinity of Fort Necessity.

After completion of my internship, I decided to do research on the CCC in West Virginia. I researched many archives and libraries, and eventually found a bibliography compiled by the Federal Security Agency on CCC literature in 1943. I wondered why such an important part of our state and federal conservation heritage had not been updated since World War II. This led me to start searching for materials produced for the CCC as well as materials about the CCC.

An opportunity arose while working in history research at the Institute for the History of Technology and Industrial Archaeology at West Virginia University. I was given the task of researching construction methodology employed by the CCC on a national basis. This work required an extensive literature search, consisting of training manuals, books, thesis and dissertations, periodicals, government documents, newspapers, films, maps and locations of CCC archives and repositories. It was then that I decided to compile my findings into a comprehensive, updated bibliography.

The research methodology involved many areas. A search of federal holdings was started. After finding the original bibliographies, I searched the Library of Congress Online Catalog for books and serials related to the CCC. I then searched the Library of Congress Prints and Photographs Division Online Catalog for their holdings. The National Register of Historic Places Online Catalog was then searched to provide the type of materials they had documented in their collection. A search was then conducted on the National Archives Information Locator (NAIL) to see what film and still image (photograph) collections were held in the National Archives. The National Park Service Libraries provided a wealth of information on training manuals, organizational structures and federal guidelines. The National Agricultural Library also provided me with a listing of references and archives held by the Department of Agriculture. The U.S. Office of Education was heavily involved in the training and education of the enrollees and it published many articles in education journals and books.

The second phase of my research involved the canvassing of all state historic preservation offices and archives. This provided holdings on a state-by-state basis, and provided me the idea of how many states had conducted an inventory of CCC structures. This phase also included a canvassing of all state parks and recreation departments for information related to the construction of CCC structures, whether it be printed materials, photographs, or drawings. The state holdings provided many leads as to the literature (books and periodicals) used in this bibliography.

The final phase of my research included the searching of books and periodicals that were used during the era of the CCC (1933-1942) and during the fiftieth anniversary exercises of the 1980s. This included searching for theses and dissertations from 1933 to the present. Federal publications and organizational journals ran many articles dealing with the progress and accomplishments of the CCC. A search of online booksellers provided access to many titles and local histories available for sale. Searching online databases such as *FirstSearch, WorldCat, Dissertation Abstracts* and others provided lists of materials written over the last several years and physically searching *Readers' Guide to Periodical Literature*, and *Education Index* provided a list of articles written from 1933 to the present.

This bibliography would not have been possible without the help of many organizations and people, even in the world of online databases and the Internet. It is necessary to first thank the Historic American Buildings Survey/Historic American Engineering Record (HABS/HAER) Division of the National Park Service. HABS/HER played a very important role in sponsoring this research. Without their support this bibliography would not have been started. Many thanks are due E. Blaine Cliver and his staff in the HABS/HAER office in Washington for their continued support.

I would also like to thank all of the numerous state historic preservation offices, archives and parks and recreation departments that responded to my questionnaire on their holdings. The materials sent, along with comments and suggestions, were very helpful in preparing this bibliography.

Thanks are also due to Dr. Emory L. Kemp and the staff at the Institute for the History of Technclogy and Industrial Archaeology, Eberly College of Arts and Sciences, West Virginia University, Morgantown, WV. Their encouragement and support in this project are responsible for its coming to fruition.

Many thanks are also due to the Interlibrary Loan Department, Charles Wise, Jr. Library, West Virginia University. The ILL staff spent many hours assisting me in locating articles and books and verifying citation information. I thank them for their patience and the time they afforded me in their busy schedules. This undertaking could have never been completed without their help.

Finally, I would like to thank all those who reviewed this bibliography. Their recommendations and suggestions were greatly appreciated. Thanks also to anyone who I have failed to mention, whether individual or institution. This bibliography was a joint effort which could not have been accomplished alone.

Larry N. Sypolt

INTRODUCTION

The Emergency Conservation Work (ECW), the original name for the agency that was to become known as the Civilian Conservation Corps (CCC), was authorized by the 73rd Congress under Public Act No. 5. The act, known as the "Reforestation and Relief Bill," was approved by President Franklin D. Roosevelt, on March 31, 1933. It gave the President authority to establish a chain of forest camps in which unemployed young men could be placed to protect and improve our nations millions of acres of forest land.[1]

The name was officially changed to the Civilian Conservation Corps by act of June 28, 1937, which established the CCC as an official agency within the federal government. The new CCC agency had a stronger emphasis on education and vocational training than the original ECW forest camp agency.

The administration of the ECW, and later CCC, was composed of a director, originally Robert Fechner, and representatives of the four cabinet agencies involved in its makeup, namely the War Department and Departments of the Interior, Agriculture, and Labor. Each department was responsible for certain activities in the coordination of the program.

The War Department was responsible for the physical maintenance of the camps and the enrollees. Under the leadership of military and naval officers, this included fiscal matters, health, supplies, shelter, transportation, communication, and cooperation from the U.S. Office of Education.

The Department of the Interior cooperated in the supervision of the CCC camps in national parks, on Indian reservations and in the territories of Alaska, Hawaii and the Virgin Islands. It also cooperated in connection with the Grazing Service, Fish and Wildlife Service, Bureau of Reclamation, and General Land Office activities.

---

[1]Ned H. Dearborn, *Once in a Lifetime: A Guide to the CCC Camp.* (New York: Charles E. Merrill Company, 1936), 5.

The Department of Agriculture cooperated through the U.S. Forest Service, Soil Conservation Service, drainage, Beltsville Research Station, plant industry, entomology, and plant quarantine programs. The Department of Agriculture was also responsible for work done on private lands and in state forests.

The Department of Labor was responsible for the selection of enrollees, upon recommendations of the state relief agencies. Need and population quotas determined who and how many were to be selected.

The organization of the CCC was divided into junior enrollees, war veterans (chosen by the Veterans Administration), Indians, blacks, and other groups. Each company consisted of a commanding officer, executive officer, and staff doctor who was provided by the War Department, and enrolled "overhead" who were not regular members of the work section. The overhead consisted of leaders, assistant leaders, company clerks, storekeepers, supply officers, infirmary attendants, stewards, cooks, chauffeurs, mechanics and assistant educational advisors. The work sections were divided into crews under foremen and assistant foremen.[2]

Each camp had a letter designation which indicated what type of camp it was, and under what jurisdiction it fell. These camp designations or symbols were as follows:

A--Agriculture (Bureau of Animal Industry); Forest Service and Bureau of Animal Husbandry; Department of Agriculture.
Army--Military Reservations: U.S. Army, War Department.
BF--Federal Game Refuge (Biological Survey); Bureau of Biological Survey; Department of Agriculture.
BR--Bureau of Reclamation; Department of the Interior.
BS--Biological Survey
C of E--State Land (Corps of Engineers); U.S. Army; War Department.
CP--County Park
D--Private Land (Drainage); Bureau of Agricultural Engineering; Department of Agriculture.
DG--Public Domain; Division of Grazing; Department of the Interior.
F--National Forests; U.S. Forest Service; Department of Agriculture.
GLO--General Land Office.
L--State and Federal Land (Levee); U.S. Forest Service and State; Department of Agriculture.
MA--Metropolitan Area Municipal Park; National Park Service.
MC--Private Land (Mosquito Control); U.S. Forest Service and State; Department of Agriculture.
MP--Military Park; National Park Service; Department of the Interior.

---

[2]U.S. Congress. Senate. *Civilian Conservation Corps*. 77th Cong. 2nd Session, Document No. 216, 1-3.

NA--National Arboretum (Bureau of Plant Industries); U.S. Forest Service; Department of Agriculture.
Navy--Naval Military Reservation; U.S. Forest Service and U.S. Navy (except Navy-1-VA, by National Park Service and Navy); Department of Agriculture.
NHP--National Historical Park; National Park Service, Department of the Interior.
NM--National Monument; National Park Service, Department of the Interior.
P--Private Forest; U.S. Forest Service and State; Department of Agriculture.
PE--Private Land Erosion; U.S. Forest Service and State; Department of Agriculture.
S--State Forest; U.S. Forest Service and State; Department of Agriculture.
SCS--Private Land; Soil Conservation Service; Department of Agriculture.
SP--State Park; State Park Division of National Park Service; Department of the Interior.
TVA--Tennessee Valley Authority; U.S. Forest Service and Tennessee Valley Authority; Department of Agriculture.
TVA-P--Tennessee Valley Authority; State Park Division of National Park Service and Tennessee Valley Authority; Department of the Interior.[3]

It is necessary to know the make-up of the CCC administration and the individual camps when researching them. Record locations vary, and the camp designator should lead the researcher to the right source.

The administration of the CCC was also divided into nine corps areas. The army, which administered the program, was already broken down into nine administrative corps areas, and these were retained by the CCC. The state's corps area jurisdiction and the headquarters location are as follows:

First--Maine, New Hampshire, Vermont, Massachusetts, Rhode Island, Connecticut. HQ: Army Base, Boston, Massachusetts.
Second--New Jersey, Delaware, New York. HQ: Governors Island, New York
Third--Pennsylvania, Maryland, Virginia, District of Columbia. HQ: U.S. Post Office and Court House, Baltimore, Maryland.

---

[3] *Happy Days CCC Directory* (Washington: Happy Days Publishing Company, 1937), 3; and, Colorado State Archives, Colorado Civilian Conservation Corps Records at the Colorado State Archives web: http://governor.co.us/gov_dir/gss/archives/.

Fourth--North Carolina, South Carolina, Georgia, Florida, Alabama, Tennessee, Mississippi, Louisiana. HQ: Post Office Bldg., Atlanta, Georgia.
Fifth--Ohio, West Virginia, Indiana, Kentucky. HQ: Ft. Hayes, Columbus, Ohio.
Sixth--Illinois, Michigan, Wisconsin. HQ: U.S. Post Office Bldg., Chicago, Illinois.
Seventh--Kansas, Arkansas, Iowa, Nebraska, Minnesota, North Dakota, South Dakota, Missouri. HQ: Federal Bldg., Omaha, Nebraska.
Eighth--Texas, Oklahoma, Colorado, New Mexico, Arizona, Wyoming (less Yellowstone National Park). HQ: Ft. Sam Houston, San Antonio, Texas.
Ninth--Washington, Oregon, Idaho, Montana, Utah, Nevada, California, Yellowstone National Park (in Wyoming). HQ: Presidio of San Francisco, California.[4]
(A map of the Corps Areas appears on page 8).[5]

Each corps area was subdivided into administrative districts. Each had a district commanding officer, who along with his staff saw to the needs of each CCC camp in his district.

This administrative hierarchy helps one understand the complex composition of the Civilian Conservation Corps. The researcher needs to know the corps area, and the agency which administered a camp to locate pertinent federal records in the National Archives. The Franklin D. Roosevelt Presidential Library is also a good source for CCC information. The address for the FDR Library is: Franklin D. Roosevelt Library (NRL), 511 Albany Post Road, Hyde Park, NY 12538. Phone: (914) 229-8114. Fax: (914) 229-0872.

The National Archives contains a record group (RG) titled "Civilian Conservation Corps" that includes many CCC records. However, records can also be found in the record groups of the National Park Service, Soil Conservation Service, Department of the Interior, Department of Agriculture, War Department, etc. A list of related record groups and their numbers are listed in the "Archive and Repository" section of the bibliography.

Many other sources and techniques for researching CCC history exist. This guide and bibliography will list general, historic, educational, governmental,

---

[4]*Happy Days CCC Directory*, 5.

[5]Illustration by Paul M. Boxley, Institute for the History of Technology and Industrial Archaeology, West Virginia University, Morgantown, WV, December 1999, and Dearborn, *Once in a Lifetime*, 253.

work projects and other references separated under headings by the Corps Area in which the material cited fits geographically.

The bibliography will be the most comprehensive one compiled in more than fifty years. It is not complete, as no bibliography ever is. It is not possible, even in this day of electronic computer retrieval, to find every known reference to the CCC. For example, thousands of local newspapers across the nation carried stories of local CCC activities. Many of these newspapers were never indexed, so listing them and their references would be an insurmountable task. The same would be true for many local and regional journals and local histories. When researching a specific camp or state, it is suggested that the researcher first look into local and state sources before going to the National Archives or Library of Congress.

Textual Resources

There are many starting points when researching the CCC. The best would be the card catalog at the local library, as many printed works were produced about and by the CCC. Start your search under "Civilian Conservation Corps" and "Emergency Conservation Work," the logical title, subject, author or keyword headings.

Larger libraries may hold the *National Union Catalog*, a multi-volume compendium of the holdings of libraries around the country. Arranged by author, subject and title, the *National Union Catalog* was first published in 1956, and has been updated periodically. It will provide the researcher a starting point, if the library is not automated or does not have access to outside online catalogs.

The *Monthly Catalog of United States Government Publications* is a good source for finding what publications were produced by the federal government. For the period 1933-1942, publications were listed under the agency or department producing them, and can be searched that way. The *Monthly Catalog* lists "see also" cross-references, where other related material can be found, under different headings. It is also possible for a publication to be listed under the producing agency heading and the CCC heading, as a subject entry.

The *Catalog of the Public Documents of the Seventy-Third Congress and of all Departments of the Government of the United States*, commonly known as the *Documents Catalog*, indexes each department of the federal government by Congress. Hence, the 73rd Congress (1933-34) would be indexed in Volume 22, the 74th Congress (1935-36) in Volume 23, the 75th Congress (1937-38) in Volume 24, and the 76th Congress (1939-40) in Volume 25. It is necessary to search under both "Emergency Conservation Work" and "Civilian Conservation Corps," as well as the "see" and "see also" headings. Listings are also found under "C.C.C. Camps" and the federal agency or department that is responsible for the publication.

For information on CCC companies and camps in a local area, local newspapers are a good source of information. Almost all newspapers carried articles of interest about the activities of the enrollees and progress of local work projects. Many newspapers from this period are not indexed, but are an excellent source of information. CCC information is usually carried in the front page or on the standard local information section.

State archives and state parks and recreation departments will generally have information about CCC activities. Many state archives have received donations of CCC materials and oral histories over the years. Camp newspapers, photograph collections, manuscript materials and other kinds of memorabilia are housed in state archives and history collections. A simple letter of inquiry might produce unexpected results.

Similarly, state parks and recreation departments have information on the formation and construction of projects which were done in the park systems. State park projects were done by CCC companies under the jurisdiction of the Department of the Interior, National Park Service. State forest projects were administered by the Department of Agriculture, U.S. Forest Service. Both agencies had standard drawings and blueprints of various types of buildings, along with construction manuals on how to build many types of structures. State parks and recreation offices as well as the state park or forest headquarters might contain various types of construction information and photographs in their holdings.

The Historical Records Survey (HRS) was created during the New Deal to inventory state archives and historical manuscript collections. Originally part of the Federal Emergency Relief Administration, HRS became an independent program in 1936. HRS documented state and local records and inventoried, cataloged and analyzed public materials, and published them as state inventories. If prepared for the state being researched, these inventories are a very valuable resource.

Another good source of written information is the historic manuscripts produced by the Federal Writers' Project, under the Works Progress Administration. They assisted in similar works as the Historical Records Survey.

The National Association of Civilian Conservation Corps Alumni (NACCCA) is also a valuable source for information. NACCCA holds various types of CCC materials including photographs, books, manuscript collections, printed company and district histories, oral histories, reports, correspondence and artifacts dealing with the CCC. They can be reached at NACCCA, P.O. Box 16429, (16 Hancock Avenue, Jefferson Barracks), St. Louis, MO 63125-0429. Phone: (314) 487-8666, Fax: (314) 487-9488. The current archivist, Harry Dallas, is very helpful in assisting researchers locate camps and other types of records. Membership dues in the NACCCA are $20.00/yr. or $100.00/life membership. NACCCA publishes a monthly paper, *NACCCA Journal,* which contains human interest articles submitted by members and former CCCers, announcements of reunions, and various other CCC items of interest. The NACCCA archives can be contacted by mail, phone, fax or E-mail: naccca@aol.com. A web page has been established which can be viewed at: http://www.cccalumni.org/. The archivist is a part-time volunteer, so allow ample time for a response.

A search for personnel records of anyone who served in the CCC would be made through the National Personnel Records Center, 111 Winnebago Street, St.

Louis, MO 63118-4199. This source is only available to the next-of-kin. When writing for information, please provide as much information as possible on the person you are researching. No phone requests are accepted.

The library and archival holdings of the National Park Service library at Harpers Ferry, WV, contain a wealth of information. The library holdings are online internally, but not yet accessible through the web or Internet. A wide variety of publications are available through the library, as well as its vast oral history collection. Over a dozen manuscript boxes contain a wide variety of pamphlets, correspondence, oral histories, and a complete set of the CCC *Project Training (P.T.)* manual series. The list of *P.T.* manuals can be found under the federal documents section of this bibliography, under the Civilian Conservation Corps entry. Among the CCC oral histories are those of S. Herbert Evison, Newton Bishop Drury, Fred Arnold, Ed Hummel, James F. Kieley, and Conrad Wirth.

The NPS Conservation Center at Harpers Ferry is also a good source for photographs and various CCC materials. The Center holds numerous historic and contemporary photographs, negatives, slides and transparencies from across the country, dealing with all types of CCC activities. Photographs exist on state and national work projects, demonstration projects and the Los Angeles Exposition of 1935.

Current information for the Harpers Ferry Center is:

Mr. David Nathanson, Chief, Library and Archival Services, Harpers Ferry Center, P.O. Box 50, Harpers Ferry, WV 25425. Phone: (304) 535-6262 and Fax: (304) 535-6492.

Mr. Tom Durant, Photo Librarian, Harpers Ferry Center, P.O. Box 50, Harpers Ferry, WV 25425. Phone: (304) 535-6707 and Fax: (304) 535-6712.

Early Bibliographies

There were bibliographies compiled during the existence of the Civilian Conservation Corps. The first of these was done by the Library of Congress, Division of Bibliography, in March 1936. There were two supplements to the original work, in June 1937 and May 1939, respectively. Two later bibliographies were done by the Office of the Director, Civilian Conservation Corps in 1939 and 1943.

These early bibliographies list many original resources. They are very helpful as a starting point for research on the literature and documentation of the CCC program. They list books, articles, government documents by the CCC and sponsoring agencies, masters' theses and doctoral dissertations. They are as complete as they could be at the time they were compiled, but further searching in other printed sources is necessary to find the full range of available articles.

A good accompanying reference source to the bibliographies is the *Reader's Guide to Periodic Literature*, which lists many articles that are not in the published bibliographies. Remember that the 1939 and 1943 CCC bibliographies were not prepared by the Library of Congress, but by the Office of the Director,

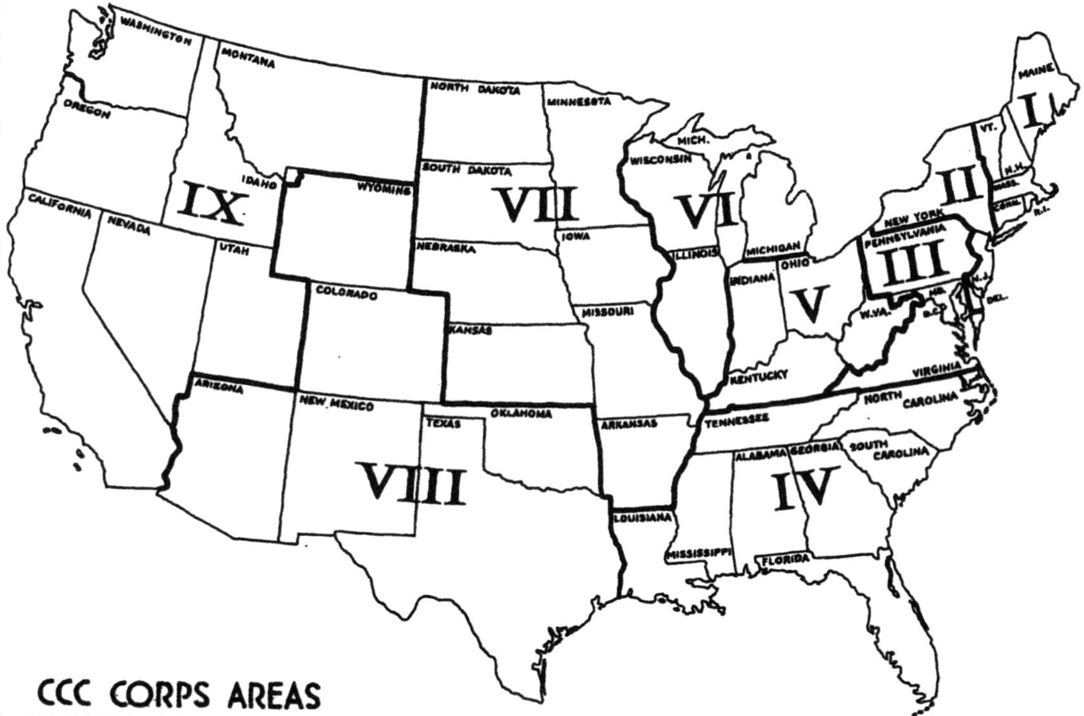

CCC CORPS AREAS

ILLUSTRATED BY: PAUL M. BOXLEY, IHTIA
DECEMBER 1999

Civilian Conservation Corps, who might not have had access to the vast LC network of resources.

There are many periodicals that produced annual indexes to their respective publication. A few of these, *American Forests*, *Forestry News Digest*, *The Regional Review*, and *Reclamation Era* as well as indexes, such as the *Education Index*, the *Engineering Index* and the *Agricultural Index* list articles by and on the CCC. *Reclamation Era* published articles almost monthly, from April 1936, on CCC activities.

The *Education Index* lists many articles on the CCC. Howard W. Oxley served as director of the CCC education program, and many of his articles are listed there. Oxley's articles appeared regularly in *School Life* and other education journals.

The CCC also had a national newspaper. *Happy Days* appeared almost as quickly as the CCC itself, was published weekly, and dealt strictly with the activities of the CCC. There was a page dedicated to news from each of the nine corps areas, and national CCC news appeared on the front page. *Happy Days* encouraged enrollees to become salesmen, and advertised many products, such as CCC jewelry, pillow covers, pennants and various other sundries. Sales and writing competitions were held by *Happy Days* which included "Best Camp Paper" and "Best Salesman." It had a positive effect on the enrollees and gave them news and information about CCC camps in other parts of the country.

The *Annual Reports of the Director*, Civilian Conservation Corps, contain a wealth of information, although generally national in nature. The *Annual Reports* start with a preliminary report on June 30, 1933 and continue through the final report of June 30, 1942. A Final Report of the CCC is published in volume No. 10676, of the *Government Documents Serial Set*. It is simply titled *Civilian Conservation Corps*, Senate Document No. 216, 77th Congress, 2nd Session.

Electronic Resources

Information retrieval has progressed a long way over the last few years. Information that was once only available by a letter or phone call is now available almost instantly by computer over the World Wide Web (web). The web contains many search engines that provide a host of indexes and locators. Any subject can be searched by simply logging on to your local network provider and typing the search engine address (URL) at the location bar.

FirstSearch is a database that contains a large amount of information on the CCC. It is only available at libraries that are members of Online Computer Library Center (OCLC). If the researcher has access to a member library, he or she should access the *FirstSearch* database and then choose "All Databases." At the next screen, choose the WorldCat database, which records any type of material cataloged by OCLC member libraries from the 12th century to present. When typing "Civilian Conservation Corps" in the "Search For" box, more than eleven hundred hits will be produced. These will be a variety of short titles for books, articles, photographs, oral histories, camp newspapers and other related materials. By clicking on the record title, the full entry will be displayed, including the holding location. Items may be ordered through Interlibrary Loan (ILL) for the researcher by following the steps on the next few screens.

One of the most valuable CCC reference works is *The Civilian Conservation Corps Camp Papers: A Guide*, by Marlys Rudeen, at The Center for Research Libraries (CRL) in Chicago. The CRL home page is online at: http://wwwcrl.uchicago.edu . The CCC database is available at: http://wwwcrl.uchicago.edu/DBSearch/Search.html. If going to the home page, the researcher should choose "Search CRL Databases." The first one to appear on the next screen is the CCC database. Here the researcher can search for CCC camp newspapers by paper title, CCC company number, company type (N, P, SCS, etc.), city, state or county. It is also possible to get an alphabetical or numerical listing of all companies in a state or county.

There are search engines that search other search engines such as Google and Dogpile. Their addresses are: http://www.google.com and http://www.dogpile.com respectively. Just type the words "civilian conservation corps" in the search box, and the search will locate thousands of sites.

While a great many are true hits for searching "Civilian Conservation Corps," many are searches that hit one or more of the words. You should browse all to ascertain their significance to the subject. The hits will include books, camp newspapers, work projects, museums, photograph collections, oral histories, federal documents, discussion lists and other information dealing with the Civilian Conservation Corps. The researcher will need to determine which hits pertain to his/her research, and pursue those avenues.

There is also a newly created New Deal Network on the web. This network can be an important research tool. The New Deal Network will put the researcher in contact with New Deal era activities. Research questions appear frequently on the network, and the information is shared by all subscribers. To my knowledge there is currently no Listserv specifically dedicated to the CCC. The address to subscribe to the New Deal Network is: http://newdeal.feri.org/. Then follow the instructions for obtaining CCC information.

There is also a Listserv for the New Deal. The address is NEWDEAL@LISTSERV.UIC.EDU. Send an E-mail message to that address and in the subject line put the word "subscribe." In the message area put the message "subscribe NewDeal Listserv," with your E-mail address. At the time of this writing, there were more than two hundred members of this Listserv and, although not specifically CCC oriented, it is a good Listserv from the era.

Federal Government Online Information

The federal government provides several information locators online to anyone who has access to the web. Once thought to be difficult for researchers, the LC, National Archives (NARA) and all departments of the federal government provide home pages with instructions on how to access their holdings and other governmental information.

The Library of Congress can be searched for materials on the CCC by accessing their web address at: http://www.lcweb.loc.gov and follow the directions to their online holdings. Page down to "Research Tools" and choose Library of Congress Catalogs. Then choose "Browse Search." Your choices are "Books Cataloged Since 1975," "Books Cataloged Since 1898," "Serials" and "Maps." All of these will provide information when the words "Civilian Conservation Corps" are typed in the subject line.

Any book in the Library of Congress can be searched by doing a title, author, subject or keyword search. A subject or keyword search for "Civilian Conservation Corps" should be the most productive. The items may then be checked for at the local library or requested on an interlibrary loan from a lending library by the local library for the researcher. Libraries normally charge a small fee (usually mailing cost) for using interlibrary loan services.

The serials browser will mostly produce camp newspapers. The serials browser shows more than one hundred alphabetical items. The map browser only offers three items but one is a map of the United States showing all of the Soil Conservation Corps CCC camps from 1933-1942.

The National Archives and Records Administration, most often referred to as the National Archives, can be a little more complex. It is very easy to access the databases of the National Archives and its Regional Branches, but the information provided on the home pages is general in nature. Collectively, the National Archives contain millions of linear feet of records from all departments of government. The records are broken down into Record Groups (RG). Each Record Group is broken down further into Sub-Groups, then smaller levels of classification called Series. It is necessary to contact an archivist at the location you want to visit before the trip is made. Much confusion can be avoided by verifying the location of the information being sought.

Civilian Conservation Corps records in the National Archives will not always be found in the most obvious place. The Record Group for the Civilian Conservation Corps is RG 35. There were, however, several other government agencies involved in the administration of the CCC. Each has its own Record Group number and should be investigated along with RG 35.

For example, the index for the Mid-Atlantic Region of the National Archives lists Civilian Conservation Corps in the Record Groups for the National Park Service (RG 79), Forestry Service (RG 95) and Soil Conservation Service (RG 114), but nothing under Civilian Conservation Corps (RG 35). Since these federal agencies were technically in charge of the work projects, their record groups contained the recorded information.

Further research will often uncover additional information. Reference to the CCC's clearance work was found in the Corps of Engineers (RG 77) files. They had not been indexed because it was an insignificant part of the overall series. It provided significant information, however, because it named the landowner who leased the CCC Camp ground and the terms and conditions of the lease. It pays to look in the not so obvious place.

The Library of Congress, Prints and Photographs Division is also an excellent place to check for CCC materials. It can be accessed at: http://lcweb2.loc.gov/pp/pphome.html . A recent search of "all catalogs" on the Prints and Photographs Division home page resulted in 78 hits. These ranged from simple photograph collections to Historic American Buildings Survey/Historic American Engineering Record (HABS/HAER) documentation. The HABS/HAER database will eventually provide full text searching, with photograph and measured drawing retrieval available from any remote site. HABS/HAER is a division of the National Park Service. The Historic American Buildings Survey (HABS) is the oldest surviving agency of the New Deal era.

The National Register of Historic Places Division of the National Park Service also contains documentation on the CCC. The National Register can be found through the NPS homepage at: http://www.cr.nps.gov/ , or by going directly to the National Register Information System (NRIS) listing page at: http://www.cr.nps.gov/nrishome.htm . Choose the Telnet connection, and log in with "NATREG." Progress through a series of information screens using "conservation" as the word to be searched. Only one keyword is allowed for searching. Browse the entries that have the words "civilian conservation corps." Only twelve (12) entries will appear on each screen, but you can proceed to the following pages by hitting the "enter" key. This database is sometimes difficult to search through Telnet, but it contains information on CCC buildings, camps and projects that have been entered on the National Register.

The *National Union Catalog of Manuscript Collections* (NUCMC) is also online. NUCMC was originally only available in hard copy from 1959 through 1994. The hard copy was then discontinued when the electronic version went online.

NUCMC is a free cooperative cataloging program provided by the Library of Congress to repositories who do not have access to the Research Libraries Information Network (RLIN), a national-level database. NUCMC catalogers create Machine Readable Cataloging (MARC) bibliographic records in RLIN, describing collections throughout the nation, and establish appropriate name and subject authority headings for the collections. The network makes descriptions and locations of the materials available to researchers.

Located at http://lcweb.loc.gov/col/nucmc/nucmc.html , a search in NUCMC for "Civilian Conservation Corps" produced 156 records. These records contained all sorts of materials including oral histories, personal paper collections, camp newspapers, manuscript collections and photographs. Clicking on the individual record produces the full cataloging entry, including where that particular collection is housed.

NUCMC has another special feature available under their web heading. Located at: http://lcweb.loc.gov/coll/nucmc/other.html is "Archival and Manuscript Repositories in the United States." This location provides an alphabetical, state-by-state listing of archival repositories, including addresses and telephone numbers. Other information about the repositories holdings is also available. In most instances, the listing for the state archives is also available.

The National Archives Information Locator (NAIL) is another means of electronically searching records held in the National Archives. The National Archives can be accessed at: http://www.nara.gov . Page down to "Research Tools" and press enter. Choose NARA Archives Information Locator (NAIL) and press enter. On the search line, type Civilian Conservation Corps, and click on "Submit Search."

One NAIL search produced more than 200 hits. These were primarily photographs, motion pictures and some textual records, all dealing with the CCC. By selecting the "Full Display" after each entry, the researcher gets the full record, with catalog number.

BIBLIOGRAPHY

GENERAL WORKS

1. Alexander, Thomas W. "Comments on the C.C.C." *Journal of Forestry* 34 (March 1936), 292-7.

2. Allen, Frederick Lewis. *Since Yesterday: The Thirties in America.* New York: Harper and Row, 1939.

3. "America Views a Permanent C.C.C." *Conservation* 1 (October 1935), 1.

   This article contains the opinions of many newspapers from around the country. All those excerpted remark about the good work of the CCC, and hope its termination is only due to the lack of unemployed youth who need the benefits of employment.

4. Amidon, B. "Future of CCC and NYA." *Survey* 77 (December 1941), 351-2.

5. Anderson, Floyd. "Unemployment and the CCC." *America* 52 (March 16, 1935), 538.

6. Armstrong, Frank H. "Civilian Conservation Corps Revival." *Journal of Forestry* 69 (April 1971), 224-5.

7. "Artists to Make C.C.C. Immortal." *American Forests* 40 (April 1934), 183.

8. Austin, C.M. "The CCC with the Navy." *U.S. Naval Institute Proceedings* 61 (June 1935), 781-90.

9. Babcock, John Gilbert Chittenden. "The Role of Public Discourse in the Soil Conservation Movement, 1865-1935." Ph.D. diss., University of Michigan, 1985.

10. Baker, Richard S. "Forest Camps of America." *Military Gazette* 50 (June 25, 1935), 6-8.

11. Baldridge, Kenneth W. "Reclamation Work of the Civilian Conservation Corps, 1933-1942." *Utah Historical Quarterly* 39 (Summer 1971), 265-85.

12. Barry, Arlene. "Is the Civilian Conservation Corps of the 1930s A 1990s Approach to Dropouts and Illiteracy?" *Journal of Adolescent and Adult Literacy* 42:8 (May 1999), 648-59.

This article discusses the standards of literacy training in the CCC camps of the 1930s, and how they taught basic reading and writing skills to enrollees. The *Enrollees Handbook* and other period workbooks and methods are looked at with today's problem in mind.

13. Bass, Thomas A. "A Reborn CCC Shapes Young Lives With an Old Idea." *Smithsonian* 14 (April 1983), 57-66.

14. Beam, Russell A. "Contributions of the Civilian Conservation Corps to Character Development." *Religious Education* 32 (January 1937), 43-9.

15. Bennett, H.H. "Sunshine and Shadow." *Soil Conservation* 3 (August 1937), 30-3.

16. Best, Gary D. *FDR and the Bonus Marchers*. Westport, CT: Praeger, 1992.

This book is about the rights of World War I veterans. Veterans were eventually permitted to be in special camps, even though some were married.

17. "Birth of the CCC." *American Forests* 79 (March 1973).

18. Bishop, Freeman C. "The Civilian Conservation Corps Carries On." *American Forests* 39 (November 1933), 507-9.

19. _____. "Remembered Men: Civilian Conservation Corps." *American Forests* 39 (August 1933), 350-2.

20. _____. "Six Months With the C.C.C." *Evening Star*, Washington, D.C. (October 11, 1933), A-3; (October 12, 1933), A-4; (October 13, 1933), A-4; (October 14, 1933), B-5; (October 15, 1933), B-4; (October 16, 1933), A-5.

21. Black, Charles Ira. "A Comprehensive Study of the Civilian Conservation Corps (1933-1942) and the Job Corps (1965-)." Master's thesis, University of Maryland, 1966.

22. Blair, T.B. "How the CCC Has Paid Off." *American Forests* 60 (February 1954), 28-30.

23. Blocker, R. "Mr. Roosevelt's Civilian Conservation Corps." *Yankee* (August 1977), 66-71 and 130-133.

24. Bogle, Lori Lyn. "Creating an American Will: Evangelical, Democracy and National Security, 1913-1964." Ph.D. diss., University of Arkansas, 1997.

25. Bowen, Doris Omundson. "America's Greatest Conservation Army." *Stronghold* 3rd ed. (July 1990), 1+.

26. Braeman, John, Robert H. Bremner, and David Brody, eds. *The New Deal: The National Level* (vol. 1). Columbus, OH: Ohio State University Press, 1975.

The subject of the CCC is scattered throughout this volume on the New Deal at the national level. There are sections on the connection of the CCC with other relief agencies, the Negro and the New Deal and the political decisions to do away with recovery programs when World War II begins.

27. _____. *The New Deal: The State and Local Levels* (vol. 2). Columbus, OH: Ohio State University Press, 1975.

The CCC is looked at through different programs and socioeconomic conditions in various states. Also discussed are other federal relief agencies in various states and territories.

28. Bramble, W.C. "Forestry and the C.C.C." *Minnesota Conservationist* 16 (September 1934), 6-7.

29. Brant, Irving. *Adventures in Conservation with Franklin D. Roosevelt.* Flagstaff, AZ: Northland Publishing, 1988.

30. Bremer, William W. "Along the American Way: The New Deal's Work Relief Program for the Unemployed." *The Journal of American History* 62 (December 1975), 636-652.

31. Bromley, Dorothy E. "The Forest Army That Lives By Work." *New York Times*, July 23, 1933.

32. Broughton, C.E. "Building for the Future." *Forestry News Digest* (February 1936), 18.

33. Brown, George Sterling. "The Civilian Conservation Corps Program in the United States." *Empire Forester* (1934), 19-32.

34. _____. "Forests and Men Benefitted by C.C.C." *New York Times* (October 8, 1933), sec. 9, 2.

35. Brown, G. Stewart. "CCC Camps as Red Cross Evacuee Centers." *American City* 57 (January 1942), 37-8.

This article tells of an agreement between the Red Cross and the CCC to train key personnel and utilize CCC camps and equipment if necessary in the upcoming national emergency. It also says that certain CCC camps would be used as evacuee centers if mass evacuations were necessitated by enemy actions.

36. Brown, Nelson C. "When East Goes West." *American Forests* 39 (November 1933), 500-1+.

37. Brown, Nelson Courtlandt. *The Civilian Conservation Corps Program in the United States*. Washington: GPO, 1934.

38. Buckles, Harold H. "I Have Learned to Know the Human Race." *American Forests* 40 (May 1934), 218.

39. Burke, Robert E. and Richard Lowitt, comps. *The New Era and the New Deal, 1930-1940*. Arlington Heights, IL: Harlan Davidson, 1941.

40. Burns, Anna C. "Esprit d'Corps: Remembering the CCC." *Forests and People* (1st Qtr. 1983), 10-14.

41. Bustard, Bruce I. "A New Deal for the Arts." *Prologue* 29:1 (1997), 62-9.

42. Butch, Wesley A. "Michigan's Timber: The Early Years, the Big Cut, and the Forest Service; A History of Resource Use and its Implications." Undergraduate Research Paper, Albion College, 1991.

43. Butler, Ovid, ed. *American Conservation in Pictures and in Story.* Washington, DC: American Forestry Association, 1935.

Butler says that the CCC is the most extensive educational project ever undertaken in forestry. Photographs throughout the article.

44. _____. "The Editor's Log [re: CCC]." *American Forests* 48 (February 1942), 53.

45. California. State Relief Administration. *Civilian Conservation Corps: Instructions, Suggestions, Questions and Answers.* [Sacramento, CA: California State Relief Administration], 1936.

46. Carr, W.G. "Federal Funds, Local Control; Reply to M. Frakes." *Christian Century* 58 (December 24, 1941), 1613.

47. Cassity, Michael J. "Huey Long: Barometer of Reform in the New Deal." *South Atlantic Quarterly* 72:2 (1973), 255-69.

48. Chan, Loren B. "California During the Early 1930s: The Administration of Governor James Rolph, Jr., 1931-1934." *Southern California Quarterly* 63:3 (1981), 262-82.

49. Chapman, H.H. "Political Activities in the Civilian Conservation Corps." *Journal of Forestry* 32 (April 1934), 397-9.

50. Christian, John L. "America's Peace Army. There is no Basis for Fear that the CCC Will Follow a Military Pattern." *Current History* 49 (January 1939), 43-45.

51. Christian, Rena. "Where Boys Become Men." *Christian Science Monitor Weekly Magazine Section* (October 9, 1935), 8-9.

52. Cilbrith, Tallie. "The CCC: The Road to Recovery." *Foxfire* 16 (Winter 1982), 228-33.

This article gives insight into the CCC in South Carolina and Georgia. It looks at organization, daily work and goals of the CCC.

53. "The Civilian Conservation Corps" *American Forests* 41 (September 1935), 530-33.

This article gives a brief history of the CCC. It also contains photos of different forest work projects.

54. "CCC and VCC Camps Build Morale." *Brotherhood of Locomotive Firemen and Engineman's Magazine* 99 (October 1935), 218-9.

55. "The C.C.C. Begins A New Year." *American Forests* 40 (April 1934), 178.

56. "CCC Begins Liquidation." *American Forests* 48 (August 1942), 379.

This story tells that all CCC camps were being closed and extra personnel would have eveything permanently shut down by August 1, 1942. It also mentions that CCC officials met with War Department, Navy Department and Civil Aeronautics Board officials to see about transfer of equipment.

57. "C.C.C. Boys are Finding Private Employment." *American Forests* 42 (September 1936), 420.

This short article tells that 145,531 enrollees left the CCC for outside employment in the fiscal year that ended June 30, 1936.

58. "CCC Camps and Transient Centers." *Wilson Bulletin* 8 (May 1934), 27+.

59. "CCC Discussion in 'Topics of the Day.'" *Literary Digest* 121 (March 28, 1936), 5.

60. "CCC: Doubles to go on Two Years." *Forestry News Digest* (May 1935), 11.

61. "C.C.C. Forest Work Starts on Second Big Year: National Forests, the Parks and State Lands Sharing in Program." *Forestry News Digest* (April 1934), 1-3.

62. "C.C.C. Forestry Work Not Detrimental to Wild Life." *American Forests* 39 (November 1933), 514.

63. "CCC: Least Criticized New Deal Unit." *Literary Digest* 121 (April 18, 1936), 48.

64. "CCC Observes Fourth Anniversary." *American Forests* 43 (May 1937), 249.

This fourth anniversary report summarizes the accomplishments of the CCC during Roosevelt's first four years. It also compliments Robert Fechner for his leadership role in the CCC.

65. "CCC Observes Seventh Anniversary." *American Forests* 46 (May 1940), 223.

This seventh anniversary report provides an update on the projects of the CCC. It also tells of the celebration held in Washington, DC, and the speakers who attended.

66. "C.C.C. On Way to Full Strength." *American Forests* 41 (July 1935), 338.

67. "The CCC Reports." *American Forests* 47 (March 1941), 133.

This article contains a general overview of the 1940 Annual Report of the Director of the CCC.

68. "The C.C.C. Speaks." *American Forests* 40 (January 1934), 21-2.

Several enrollees give testimony of their experiences in the CCC.

69. "C.C.C. Symposium." *Journal of Forestry* 32 (December 1934), 930-51.

70. "The CCC Takes to the Sea." *American Forests* 46 (August 1940), 365.

Nine hundred men were chosen from the CCC for merchant ship training. This was done under the U.S. Coast Guard and Maritime Commission.

71. "CCC Work is of 'Inestimable Value' to Future Generations, Robert Fechner, Director, Reports to President." *Commercial and Financial Chronicle* 138 (February 17, 1934), 1172.

72. "CCC Work Valued at $292,000,000." *American Forests* 41 (February 1935), 84.

73. "C.C.C. Work Valued at $579,000,000." *American Forests* 42 (February 1936), 90.

Robert Fechner tells of the accomplishments of the CCC during the program's first 30 months. It also tells the types of work projects and the amount of work completed.

74. "Civilian Conservation Corps and National Youth Administration Face Attacks in Congress; Fact Material and Pro and Con Discussion." *Congressional Digest* 21 (June 1942), 163-92.

75. "The Civilian Conservation Corps: Indispensable." *Colliers* 104 (August 19, 1939), 62.

76. "Civilian Conservation Corps May be Expanded." *American Forests* 39 (December 1933), 562.

77. "Civilian Conservation Corps Reaches Greatest Strength." *American Forests* 40 (September 1934), 420.

This informative article tells that the CCC is at its peak strength, 370,000 enrollees, in 1,648 camps nationwide. It also provides their distribution by state.

78. "The Civilian Conservation Corps Volume I." *Foxfire* 16 (Winter 1982), 226-286.

79. Clepper, Henry. "The Birth of the C.C.C." *American Forests* 78 (March 1973), 8-11.

80. _____. "Golden Anniversary of the CCC." *American Forests* 89 (April 1983), 44-5.

81. _____. "Saga of the CCC by J.D. Guthrie, Review." *Journal of Forestry* 40 (October 1942), 814.

82. Cody, Allicia. "Characteristics of 195 Civilian Conservation Corps Enrollees." Master's thesis, University of Chicago, 1938.

83. Coggins, Allen R. "The Early History of Tennessee's State Parks, 1919-1956." *Tennessee Historical Quarterly* 43:3 (Fall 1984), 295-315.

84. Cohen, Sol. "Remember the CCC?" *Phi Delta Kappan* 49 (March 1968), 369-72.

85. Cohen, Stan. *The Tree Army: A Pictorial History of the Civilian Conservation Corps, 1933-1942.* Missoula, MT: Pictorial History Publishing Company, 1993.

86. Collingwood, G.H. "Conservation Camps to Feel Political Patronage." *American Forests* 39 (September 1933), 418.

87. Conkin, Paul K. *Tomorrow A New World.* Ithaca, NY: Cornell University Press, 1959.

Primarily about the Resettlement Administration and Subsistence Homesteads, but mentions some of the work done by the CCC to help in setting up new communities.

88. Coontz, John L "Who Will Get the C.C.C.?" *Conservation* 2 (December 1936), 1.

This is an article condensed from a December 6, 1936, article in the *Washington Star* dealing with the reorganization of the CCC. It tells of the impact that the four managing agencies would have over the CCC if their agency was given sole control of the CCC.

89. Cornebise, Alfred E. *The CCC Chronicles: Camp Newspapers of the Civilian Conservation Corps, 1933-1942.* Jefferson, NC: McFarland % Co., 2004.

The author discusses the role of camp newspapers in the life of the enrolee. He explains how the enrolee lerned skills from working for the paper. He also provides the titles of many papers and how they came into being.

90. Cream, Joseph. "The Genesis of the Civilian Conservation Corps." Columbia University, 1955.

91. Culley, John Joel. "A Troublesome Presence: World War II Internment of German Sailors in New Mexico." *Prologue* 28 (Winter 1996), 278-95.

Captured German sailors form the S.S. Columbus are interred in a former New Mexico CCC camp before World War II is officially declared.

92. Cutler, Phoebe. *The Public Landscape of the New Deal.* New Haven, CT: Yale University Press, 1985.

The CCC helped landscape architects discover their country.

93. Darling, John Herrick Hiram. "The CCC Makes Good." *Christian Century* 50 (October 25, 1933), 1333-5.

94. Davis, Kingsley. *Youth in the Depression.* Chicago: University of Chicago Press, 1935.

95. Daw, T.E. ""Remember the CCC? Time Stamps Approval." As Told to Russell McKee. *Michigan Conservationist* 24 (July-August 1960), 12-4.

96. Doty, Harrison. "Our Forest Army at War." *Review of Review's* 88 (July 1933), 30-4.

97. "Doughboys of 1933 Off to the Woods." *Literary Digest* 115 (April 29, 1933), 22-3.

98. Drake, Darwood. *Looking Back.* Chicago: Adams Press, 1989.

99. "Dramatic Broadcast by Corps at 59th Annual Meeting of the American Forestry Association." *American Forests* 40 (December 1934), 576.

100. Dunham, Arthur. "Building Men, Growing Trees." *Survey* 69 (May 1933), 186-7.

101. Eaton, Warren F. "Depression Army Takes to the Woods." *Bird Lore* 37 (July 1935), 255-64.

102. Eberly, Donald, and Michael Sherraden, eds. *The Moral Equivalent of War? A Study of Non-Military Service in Nine Nations.* Westport, CT: Greenwood, 1990.

103. Edmondson, Everett L. "Some Nation-Wide Problems of the Civilian Conservation Corps." Ph.D. diss., Northwestern University, 1940.

104. "Eight Years of CCC: Training for 2,500,000 Youths Has Cost U.S. $2,779,700,000. *Newsweek* 17 (April 14, 1941), 23.

105. Ellis, Edward Robb. *A Nation in Torment: The Great American Depression, 1929-1939.* New York: Coward-McCann, 1970.

Ellis provides a very good description of what life was like during the Great Depression, and how it affected people. Several pages are given to the CCC program and some humorous experiences are mentioned. Chapter 18 deals with migrants and the CCC.

106. "Emergency Conservation Program in Pictures." *Reclamation Era* 25 (July 1935), 142.

107. "Employment Among Former Members of the Civilian Conservation Corps." *Monthly Labor Review* 40 (January 1935), 45-7.

108. "Employment Status of Former Members of the Civilian Conservation Corps." *Monthly Labor Review* 39 (August 1934), 308-10.

109. Ermentrout, Robert Allen. *Forgotten Men: The Civilian Conservation Corps*. Smithtown, NY: Exposition Press, 1982.

110. Eudy, Jesse B. *From the Depression Years A Great Need Gave Birth to the 3-C's: Saga of the CCC*. Columbia, SC: Paramount Print, 1993.

111. "The 'False Economy' of the C.C.C." *Conservation* 2 (August 1936), 4.

This article is rebuttal to the "false economy of the New Deal" remarks made by Senator Steiwer, of Oregon, at the Republican National Convention in Cleveland, Ohio. The author cites accomplishments of the CCC in the Senator's home state of Oregon, in his argument.

112. Farley, Yvonne S. "'A Good Part of Life' Remembering the Civilian Corps [sic]" and "The Civilian Conservation Corps and Babcock State Park." *Goldenseal* (January-March 1981).

113. Fear, R.A. "CCC Rebuilds Maladjusted Youth." *Phi Delta Kappan* 22 (April 1940), 385-9.

114. Fechner, Robert. "My Hopes for the CCC." *American Forests* 45 (January 1939), 10-12+.

115. Ferguson, Barbara K. "Roosevelt's Tree Army: The Civilian Conservation Corps, 1933-42." B.A. thesis, James Madison University, 1993.

116. Flynn, T. "Life in the CCC." *America* 53 (April 27, 1935), 54-5.

117. Fraine, Harold S. "The Spirit of the C.C.C." *American Forests* 40 (June 1934), 256-7+.

118. Fraine, Harold S. and Erle Kauffman. "Heroes of the CCC." *American Forests* 40 (July 1934), 303-4.

119. Fraser, Bill. *Thoughts From a CCC Chaplain*. St. Louis, MO: National Association of Civilian Conservation Corps Alumni, [1998].

A former enrollee of the CCC, Rev. Bill Fraser has compiled all of his last seven years' "Chaplain's Corner" articles, from the *NACCCA Journal*, into this little booklet. All have to do with the question, "what is our purpose in life?"

120. Friedel, Frank. *Launching the New Deal.* Boston: Little, Brown and Co., 1937.

121. Gallagher, Hugh Gregory. *FDR's Splendid Deception.* New York: Dodd, Mead & Company, 1985.

122. Gignilliat, Triece. "FDR Joins CCC Alumni for Reunion at Vogel State Park." *DNR Outdoor Report* (October 1987), 20-1.

123. Gilbertson, G.H. "The C.C.C.'s Eighth Birthday." *Soil Conservation* 6 (April 1941), 262-4.

124. _____. "CCC Marches On." *Soil Conservation* 7 (March 1942), 212-4+.

125. Gill, Tom. "Let's Look at the Record." *Forestry News Digest* (January 1934), 3.

126. Gleason, J.M. "The Building of Men in the Civilian Conservation Corps." *Northwest Science* 8 (September 1934), 16-9.

127. Goode, Thomas H. "Georgia Congressmen and the First Hundred Days of the New Deal." *Georgia Historical Quarterly* 53:2 (1969), 129-46.

128. Gorham, Eric B. "National Service, Citizenship, and Political Education." Ph.D. diss., University of Wisconsin, 1990.

129. "Government and Youth: $400,000,000 Goes to Future Citizens." *Life* 8 (April 15, 1940), 77-83.

130. Gower, Calvin W. "The CCC, The Forest Service, and Politics in Maine, 1933-1936." *New England Social Studies Bulletin* 30 (Spring 1973), 15-21.

131. _____. "'Camp William James': A New Deal Blunder?" *The New England Quarterly* 38 (December 1965), 475-93.

132. _____. "Conservation, Censorship, and Controversy in the CCC, 1930's." *Journalism Quarterly* 52:2 (1975), 277-84.

133. Green, Gilbert. "Roosevelt's 'Happy Days' for the Young Generation." *Communist* 14 (July 1935), 661-72.

134. Griffin, Harold. "Sap Among the Dingbats." *Scribner's Magazine* 97 (March 1935), 179-82.

135. Guthrie, John D. "Now They Are Men: The Story of the CCC (Review)." *Journal of Forestry* 39 (January 1941), 77.

136. _____. *Saga of the C.C.C.* Washington, DC: American Forestry Association, 1942.

Guthrie looks bact at the CCC in prose and poetry. Several drawings about CCC life, by Marshall Davis, are included.

137. _____. "School in the Camps, by F.E. Hill (Review)." *Journal of Forestry* 33 (December 1935), 1012.

138. _____. "This New America (Review)." *American Forests* 43 (May 1937), 259.

This is a review of the book *This New America* which is about the CCC.

139. _____. "Youth Rebuilds--Stories from the CCC (Review)." *Journal of Forestry* 32 (October 1934), 775-6.

140. Hanson, Val T. "When Death Rides the Forests." *American Forests* 43 (January 1937), 12-3+.

141. Happy Days Publishing Company. *Happy Days C.C.C. Directory.* Washington, DC: Happy Days Publishing Company, 1937.

These directories were published annually and listed the military personnel, educational advisor and project superintendent for each active CCC company. Also listed were all corps area and district staff.

142. Hatton, John H. "New Things in Conservation." *Producer* 15 (October 1933), 3-6.

143. Hawes, Austin F. "Can Private Owners be Induced to Practice Forestry?" *Connecticut Woodlands* 2 (February 1937), 13-14.

144. Hayden, Ernest A. *The United States Civilian Conservation Corps of the 1930s.* Callahan, CA: Ernest A. Hayden, 1985.

145. Hayward, Paul H. "Depression Army Moves to War." *Nation's Business* 21 (August 1933), 30+.

146. Helm, Douglas, comp. *Preliminary Inventory of the Records of the Civilian Conservation Corps.* Washington, DC: National Archives and Records Administration, 1980.

147. Hildreth, Richard Leroy. "A Study of the Relationship Between Golconda Civilian Conservation Corps Members' Behavior and the Phases of the Moon." Ph.D. diss., Southern Illinois University at Carbondale, 1984.

148. Hill, Edwin G. *In The Shadow of the Mountain: The Spirit of the CCC.* Pullman, WA: Washington State University Press, 1990.

149. Hill, Frank E. "CCC Marches Toward a New Destiny: Permanent Role is Seen for America's Youth Army." *New York Times Magazine* (February 21, 1937), 10-11+.

150. Hiller, Charles. "It Taught Me One Way to be Happy." *American Forests* 40 (March 1934), 121+.

151. Hilton, Ronald James. "The Short Happy Life of a Learning Society: Adult Education in America, 1930-1939." Ph.D. diss., Syracuse University, 1981.

Hilton, in one section of his dissertation, talks at length about the educational aspect of camp training. Although a youth program, there were many adults enrolled in the CCC. He also talks about the curriculum and teaching practices exercised in the camps and states that the overall educational program was "an adult education experience by both design and method."

152. Hitchcock, A.A. "Year of the C.C.C." *Clearing House* 12 (September 1937), 19-25.

153. Holland, K. "American Work Camp Movement." *Internationale Zeitschrift fur Erziehung* 6 (1937), 249-69.

154. Hosmer, Charles B., Jr. *Preservation Comes of Age: From Williamsburg to the National Trust.* 2 vols. Charlottesville, VA: University Press of Virginia, 1981.

Discusses some of the preservation projects tht the CCC were involved in throughout the country.

155. Howard, Donald S. *The WPA and Federal Relief Policy.* New York: Russell Sage Foundation, 1943.

156. Howard, Ivy M. "Pioneers of Today." *Soil Conservation* 3 (March 1938), 244-6.

157. Howell, Glenn. *CCC Boys Remember: A Pictorial History of the CCC.* Medford, OR: Klocker Printery, 1976.

This book was compiled with many stories of former enrollees of the CCC. Their personal stories are enhanced with photographs.

158. Hoyt, Ray. *We Can Take It: A Short Story of the CCC.* New York, Cincinnati, OH: American Book Company, 1935.

Hoyt explains the make up and logistics of the CCC in this early book. He tells of the organization of the four agencies involved, and how every phase of the program interacts with the other. The book also contains many illustrations by Marshall Davis.

159. Hunt, Byron. *From Alphabet Soup to WPA Nuts.* Indianapolis, IN: Maynard Miller, 1939.

160. Ickes, Harold L. *The Secret Diary of Harold L. Ickes.* New York: Simon and Schuster, 1953-54.

161. "Is It A Blind Alley?" *World Tomorrow* 17 (May 24, 1934), 270.

162. Jackson, Donald Dale. "They Were Poor, Hungry, and They Built to Last." *Smithsonian* 25 (December 1994), 66-78.

163. _____. "To The CCC: Thanks for the Memories and Monuments." *Smithsonian* 25 (December 1994),

164. Jeffers, D.S. " Further Comments on the C.C.C." *Journal of Forestry* 34 (March 1936), 298-302.

This article discusses the issues of camp work projects discipline and education of the enrollees. He questions whether the CCC is placing too much emphasis on the enrollee and not enough emphasis on the work projects.

165. Jensen, Daniel Delano. "The Civilian Conservation Corps: Youth's Second Chance." Master's thesis, San Diego State College, 1964.

166. Johnson, Frederick K. "The Civilian Conservation Corps: A New Deal for Youth." *Minnesota History* 48 (Fall 1983), 295-302.

167. Johnson, J.C. "Sailing the Three C's." *Infantry Journal* 43 (March-April 1936), 125-29.

168. Johnson, Ray. "I Can Meet My Fellows, Man to Man." *American Forests* 40 (February 1934), 67+.

Johnson tells how he has learned to work with others while in the CCC.

169. Kauffman, Erle. "Heroes of the C.C.C." *American Forests* 40 (June 1934), 303-4+.

This article tells of heroic feats performed by enrollees while serving in the CCC.

170. _____. "'Roosevelt'--Forest Camp No. 1 in the Mountains of Virginia." *American Forests* 39 (June 1933), 251-4.

171. Keegan, Andrew R. "CCC: A Successful Job Corps, 1930s Style." *American Visions* 1 (September/October 1986), 20-5.

172. Kelly, Dwight D. "Economic Characteristics of Veteran Civilian Conservation Corps Members." Master's thesis, Indiana University, 1939.

173. Kennedy, C.B. "Bases Full, and the CCC Strikes Out." *Journal of the National Education Association* 26 (December 1937), 297.

174. Kennedy, Renwick C. "Military Interlude." *Christian Century* 50 (September 13, 1933), 1144-5.

175. _____. "Quality in the Reserve Corps." *Christian Century* 52 (February 27, 1935), 272-4.

Discusses issues about serving in the Reserve Corps and working with the CCC. Many Reserve Officers got essential training while administering a CCC camp.

176. Kerr, Edward F. "Days of the Civilian Conservation Corps." *Forests and People* 13 (1st Quarter 1972), 11-39.

177. Kidwell, James. "A Task, A Plan and Freedom." *American Forests* 40 (January 1934), 23-4.

178. Kiefer, E. Kay. *Hobnail Boots and Khaki Suits A Brief Look at the Great Depression and the Civilian Conservation Corps as Seen through the Eyes of Those who were There.* Chicago: Adams Press, 1983.

179. Kieley, James F., comp. *A Brief History of the National Park Service.* Washington, DC: Civilian Conservation Corps, 1940.

180. Knox, Frank. "The CCC: Shaper of Destinies." *Pacific Northwesterner* 36:2 (1992), 17-26.

181. Krane, D.G. "Adjustment of Youth on Relief." *High Points* 18 (March 1936), 5-15.

182. Krog, Carl E. "'Organizing the Production of Leisure:' Herbert Hoover and the Conservation Movement in the 1920s." *Wisconsin Magazine of History* 67:3 (1984), 199-218.

183. Kyvig, David E. and Mary Ann Blasio. *New Day/New Deal: A Bibliography of the Great American Depression, 1929-1941.* Westport, CT: Greenwood Press, 1988.

184. "Labor Army Takes the Field." *Literary Digest* 115 (April 115, 1933), 6.

185. Lacy, Leslie Alexander. *The Soil Soldiers: The Civilian Conservation Corps in the Great Depression.* Radnor, PA: Chilton Book Co., 1976

This is one of the best and most popular works on the CCC. It gives first-hand accounts of CCC life and activities while telling the story of the New Deal' s most successful program.

186. Lafferty, Travis. *Another Day, Another Dollar: Remembering the CCC, 1933-1942.* Oakland, CA: T. Lafferty, 1990.

187. Lange, Howard L. *The CCC: A Humanitarian Endeavor During the Great Depression.* New York: Vantage Press, 1984.

Lange tells of his experiences as a medical doctor in the CCC. He describes the activities and men who made up the company he was assigned to in Missouri.

188. Lanier, Ann Thedford. "The Civilian Conservation Corps, 1933-1942: Organizational Structure and Social Effects." Master's thesis, University of Colorado, Boulder, 1983.

189. Lanpher, Henry C. "The Civilian Conservation Corps--Its Program and Recorded Accomplishments for Unemployed Youths." Ph.D. diss., University of Chicago, 1941.

190. Lasswell, James. *Shovels and Guns: The CCC in Action.* New York: International Pamphlets, 1935.

This pamphlet takes a socialistic view of the CCC and tries to show how the administration is breaking down prevailing wage standards while preparing the nation's youth for war under the guise of the CCC.

191. Laubach, H.L. "The CCC Mirrors Youth." *New York Times* (January 5, 1936), (Magazine Section) 12-3.

192. _____. "General Tests the CCC Boy." *Literary Digest* 121 (January 25, 1936), 17.

193. Lawrence, Paul A. *Remembering the CCC.* San Anselmo, CA: PAL Press, c1983.

194. Leake, Fred E. and Ray S. Carter. *Roosevelt's Tree Army: A Brief History of the Civilian Conservation Corps, 5th Ed.* Falls Church, VA: National Association of Civilian Conservation Corps Alumni, 1983.

195. Lee, Ronald F. "Objectives and Policies of Historical Conservation." *The Regional Review* 2 (March 1939), 3-8.

196. Lenoir, Nancy Ruth. "The Origins and Establishment of the Civilian Conservation Corps." Undergraduate thesis, East Texas State University, 1965.

197. Leuchtenburg, William E. *Franklin Roosevelt and the New Deal, 1932-1940.* New York: Harper and Row, 1963.

198. Lindley, Ernest K. "Civilian Conservation Corps and Other Federal Aid Projects." *School and Society* 41 (April 27, 1935), 578-80.

199. _____. "Forest Interlude: After Their Service in the CCC is Concluded, How Shall Young Americans Find Their Jobs in our World of Today?" *Today* 2 (September 22, 1934), 14.

200. Lindsey, Julian B. "The Largest CCC Camp in the Country." *Infantry Journal* 40 (July-August 1933), 267+.

201. Lorwin, Lewis Levitzki. *Youth Work Programs: Problems and Policies.* Washington, DC: American Council on Education, 1941, also: New York: Arno Press, 1974.

202. Lujan, Joe Roy. "Dennis Chavez and the Roosevelt Era, 1933-1942." Ph.D. diss., University of New Mexico, 1987.

203. Lyman, C.K. and Bert Holtby. "Corps Conservation Center." *Journal of Forestry* 63 (May 1965), 370-1.

204. McClelland, Linda Flint. *Presenting Nature: The Historic Landscape Design of the National Park Service 1916-1942.* Washington, DC: National Park Service, 1993.

This book looks at the National Park Service policy of landscape design and the important contributions to this policy by the CCC. It tells how the CCC used local materials to build roads, trails, parks, picnic areas, demonstration areas, and other projects with the local materials that were available to them. It also looks at the development of the state park systems during the nation's greatest era of park expansion.

205. McConkey, Darel. "The Youth Movement in America's Forests." *American Forests* 40 (November 1934), 509-11+.

206. McCormack, John W. "Civilian Conservation Corps." *Congressional Record* 79 (August 17, 1935), 14052.

207. McEntee, James J. *Now They Are Men: The Story of the CCC.* Washington, DC: National Home Library Foundation, 1940.

208. _____. "Point of View of the CCC." *Frontier Democrat* 8 (December 1941), 77-9.

209. McKinney, Guy D. "Army in the Forests" *Natural History* 34 (March 1934), 141-50.

210. _____. "CCC Carries On: The Record and the Program." *New York Times* (October 28, 1934), Sec. 8, 2.

211. McMillen, James A. "Best of All, It Gave Us a Job." *American Forests* 40 (March 1934), 120-1.

212. Madsen, Arch L. "CCC Veteran Recalls Volunteer Role in Signal." *Army Commo* (Summer 1994), 57-60.

213. Margason, Omar K. *The Civilian Conservation Corps: A Personal History in the Early Thirties.* San Jose, CA: O.K. Margason, 1993.

214. Markin, Estelle. "Civilian Conservation Corps Camp Programs as Evaluated by Enrollees." Master's thesis, University of Chicago, 1936.

215. Marsh, Clarence S. "Future of the CCC." *Forum* (New York) 103 (May 1940), 283-7.

216. Martin, Robert E. "Army Recruited From Idle Men Wars Against Fire and Flood." *Popular Science Monthly* 125 (July 1934), 49-52+.

217. Mass, John Witold. "The History of the Civilian Conservation Corps." Master's thesis, University of California, 1942.

218. Matheson, J.R.D. "270,000 Young Americans." *North American* 247:1 (March 1939), 10-20.

219. Maverick, Colonel (pseud). "March 'em Up and Down." *Infantry Journal* 62 (January 1948), 17-19.

220. "Men of the Three C's: Sketches by WPA Artists." *Survey Graphic* 23 (July 1934), 332-3.

221. Merrill, Perry. *Roosevelt's Forest Army: A History of the Civilian Conservation Corps, 1933-1942.* Montpelier, VT: Perry Merrill, 1981.

This book is divided into three parts: The Civilian Conservation Corps; the Value of the CCC to the Enrollees and Their Families; and the Value of the Civilian Conservation Corps to the States. The first part tells of the work projects by agency. The second tells of the benefits to the enrollees and their families for those who participated. The last part gives a state by state account of the number of camps, enrollees and a recap of the project accomplishments.

222. Miller, C.G. "Birthday Comments on the Civilian Conservation Corps." *Education* 61 (May 1941), 573-4.

223. "Misrepresenting the CCC." *American Forests* 42 (September 1936), 413.

This editorial addresses the complaint that the CCC has degenerated into politics and boondoggling work. It argues that conservation policies and politics are required to make the program move forward. It also argues that all CCC work is a step forward for conservation.

224. Mitchell, John G. "FDR's Tree Army." *Audubon* 85 (November 1983), 82-93.

225. Mitchell, Jonathan. "Roosevelt's Tree Army." *New Republic* 83 (I, May 29, 1935), 64-6, and (II, June 12, 1935), 127-9.

This article talks about problems with the CCC, compared with the CCC goals at inception. The author cites militarism and fascism as the common accusations against the CCC and advocates the removal of CCC control from the War Department and giving more powers to the educational directors.

226. Mitchell, Paul D. "Human Conservation During the Depression: A Study of the Economic and Social Contributions of the CCC, 1933-1942." Master's thesis, University of Oklahoma, 1949.

227. Monahan, R.S. "Snow Bowl." *American Forests* 43 (January 1937), 15-20.

228. Moody, Frank Kennon. "F.D.R. and His Neighbors: A Study of the Relationships Between Franklin D. Roosevelt and the Residents of Dutchess County." Ph.D. diss., State University of New York at Albany, 1981.

229. "More About C.C.C. and E.C.W." *Anthracite Forest Protector* 7 (September-October 1933), 3-5.

230. Morey, George. "Recollections About CCC Camps During the 1930's." *Timberlines* 24 (August 1981), 82-7.

231. Morrell, Fred. "What's Ahead for the CCC?" *American Forests* 48 (January 1942), 24-6.

232. Moynihan, Daniel P. "A Conservation Corps for Today." *American Forests* 94 (September/October 1988), 14-5.

233. Muller, Herman J. "The Civilian Conservation Corps, 1933-1942." *Historical Bulletin* 28 (March 1950), 55-60.

234. Muller, Max, ed. *Youth Rebuilds; With the Volunteers of the American CCC and the English Work Camps*. Leipzig: B.G. Teubner, 1935.

235. Mulvey, Deb. *We Had Everything But Money*. Greendale, WI: Reminisce Books, 1992.

236. Murray, Fred J. "Paul Bunyan Reads the Record of the CCC." *American Forests* 43 (April 1937), 154-8+.

This article, with photographs, tells of the accomplishments of the CCC, as it would have been seen through the eyes of the legendary Paul Bunyan.

237. _____. "Paul Bunyan's Rival." *Review of Reviews and World's Work* 95 (June 1937), 44-5.

238. National Association of Civilian Conservation Corps Alumni. *Roosevelt's Tree Army: A Brief History of the Civilian Conservation Corps.* Arlington, VA: The Association, 1983.

239. Neher, Leslie I. *The C.C.C. File: General Information, the C.C.C. Experience in Virginia.* Gas City, IN: L.I. Neher, 1995.

240. Nelson, M. M. (Red). "My Involvement with the Civilian Conservation Corps." *Timberlines* 24 (August 1981), 48-51.

241. Nichols, Amy and Tammi English. "We Didn't Get Much Money, Thirty Dollars a Month. I Was Awful Proud of It, and I Never Will Forget It." *Foxfire* 23 (Spring 1989), 28-31.

242. Nixon, Edgar. *Franklin Delano Roosevelt and Conservation, 1911-1945.* Washington, DC: Government Printing Office, 1957.

243. Noble, Elizabeth. *Billions for Bullets.* New York: American League Against War and Fascism, 1937.

This pamphlet explores military spending and the growing militarization and Fascism evident in the CCC camps, ROTC (Reserve Officer Training Corps), and CMTC (Citizen's Military Training Corps), as discussed by the American League Against War and Fascism. It mentions that the United States is expanding its military and naval forces under the guise of New Deal programs such as the CCC, WPA and NIRA.

244. Nolling, Wilhelm. "Federal Programs to Provide Work and Training Experience for Unemployed Youth: A Comparative Analysis of the Civilian Conservation Corps and National Youth Administration and Related Legislative Proposals for the 1960's." Master's thesis, University of California, 1964.

245. Nolte, M. Chester, ed. *Civilian Conservation Corps: The Way We Remember It, 1933-1942: Personal Stories of Life in the CCC.* Paducah, KY: Turner, 1990.

246. O'Brien, Marjorie A. "An Evaluation of the Civilian Conservation Corps as an Educational Institution." Master's thesis, Ohio University, 1970.

247. "Off to the Woods." *Producer* 14 (May 1933), 13.

248. O'Harrow, Clifford G. "The Establishment and First Six Months of the Civilian Conservation Corps [and] Joe Hill, Wobbly Martyr of Utah." Master's thesis, Utah State University, 1968.

249. Oliver, Alfred C., Jr., and Harold M. Dudley. *This New America: The Spirit of the Civilian Conservation Corps*. New York: Longmans, Green and Company, 1937.

250. Olson, James S., ed. *Historical Dictionary of the New Deal*. Westport, CT: Greenwood Press, 1985.

Olson includes entries on the CCC, its administrators and the heads of departments that worked directly with theCCC.

251. "1,500,000 To Be Benefitted [sic] this Winter by C.C.C." *American Forests* 40 (January 1934), 38.

252. Owen, Anna Lou Reisch. *Conservation Under F.D.R.* New York: Praeger, 1983.

253. _____. "Conservation Under Franklin D. Roosevelt by Anna Lou Reisch." Ph.D. diss., University of Wisconsin, Madison, 1952.

254. Oxley, Howard W. "CCC Contributes to Human Conservation." *School Life* 22 (June 1937), 310-11.

255. _____. "Enrollees Become Better Citizens." *School Life* 22 (March 1937), 215-6.

256. _____. "Enrollees Report on CCC Experiences." *School Life* 21 (February 1936), 144.

257. _____. "For Half Million Young Americans." *School Life* 21 (September 1935), 16.

258. _____. "From CCC Camp to Employment." *School Life* 22 (February 1937), 183-4.

259. _____. "Graduate Students Study CCC Education." *School Life* 23 (May 1938), 321-2.

260. _____. "Reaching Two Million American Youth." *Journal of the National Education Association* 27 (April 1938), 110-12.

261. Pack, C.L. "Human Dividends of the C.C.C." *Review of Reviews and World's Work* 88 (October 1933), 40-1+.

262. Parker, Homer C. "Conduct of the C.C.C. Camps." *Congressional Record* 78 (January 15, 1934), 668-9.

263. Paxton, Ed. "Half Billion Dollar C.C.C. Program Working Wonders for Farmers of America." *Congressional Record* 78 (March 18, 1934), 5309-10. (Article from the Sun-Democrat, Paducah, KY)

264. "Peacetime Army." *New Republic* 74 (April 5, 1933), 202.

265. Persons, W. Frank. "Emergency Conservation Camps." *Social Service Review* 7 (September 1933), 485-7.

266. Pinkett, Harold T. "Records in the National Archives Relating to the Social Purposes and Results of the Operation of the Civilian Conservation Corps." *Social Service Review* 22 (March 1948), 46-53.

Much of the information in this article comes from Pinkett's *Preliminary Checklist of Records of the Civilian Conservation Corps*, of 1944. It looks at the social consequences in the fields of employment, training, health, safety and character development.

267. _____. "Records of a Historic Thrust for Conservation." *Prologue* 8 (Summer 1976), 77-84.

268. Plair, T.B. "How the CCC Has Paid Off." *American Forests* 60 (February 1954), 28-30, 44-5.

269. Platz, John C. "Partners in Training and Preservation." *CRM* (Cultural Resources Management) 17:1 (1994), 1-8.

This article talks shows how the Forest Service is preserving the structures that the CCC built in their forests. It illustrates a working model for a preservation workshop.

270. "Poor Young Men." *Time* 33 (February 6, 1939), 10-12.

271. Prokop, John. "Do we Need Another CCC?" *American Forests* 68 (March 1962), 21-3.

272. "Proposed New State Forests." *American Forests* 42 (January 1936), 33.

Accompanying this article is a map of the United States showing proposed land purchases and locations of the closest CCC camp.

273. Prosser, Daniel. "The New Deal Builds: Government Architecture During the New Deal." *Timeline* 9:1 (February-March 1992), 40-55.

274. Putnam, Carl L. "The CCC Experience." *Military Review* 53 (September 1973), 49-62.

275. Randolph, Jennings. "The Civilian Conservation Corps: America's Productive Investment." *Wonderful West Virginia* 47:10 (December 1983), 12-13.

276. _____. "The Civilian Conservation Corps has Proved its Worth." *Congressional Record* 79 (April 10, 1935), 5674-5.

277. _____. "Shoes for the Civilian Conservation Corps." *Congressional Record* 79 (August 15, 1935), 13812.

278. _____. "The Value of the Civilian Conservation Corps. Extension of Remarks." *Congressional Record* 78 (June 13, 1934), 11870-2.

279. Ranney, Frank. "It Has Taken Me Off the Road." *American Forests* 40 (April 1934), 163-4.

This article tells how the CCC has found work in the CCC after roaming around the country looking for work.

280. Rauch, Basil. *The History of the New Deal.* New York: Creative Age Press, Inc., 1944.

281. Rawick, George Philip. "The New Deal and Youth: The Civilian Conservation Corps, the National Youth Administration, and the American Youth Congress." Ph.D. diss., University of Wisconsin, 1957 [c. 1960].

282. Reed, William J. *The Green Guidon.* Civilian Conservation Corps, Fort Missoula District, 1930.

283. Reiman, Richard A. *The New Deal and American Youth: Ideas and Ideals in a Depression Decade.* Athens, GA: University of Georgia Press, 1992.

284. Ringland, Arthur C. "The C.C.C. in Germany." *Conservation* 2 (August 1936), 8.

The author compares the United States CCC with the Voluntary Labor Service in Germany.

285. _____. "The Patriotism of Peace." *American Forests* 40 (1934), 3-6.

286. Ritter, Edward. "Memorial of Fire Heroes." *American Forests* 54 (September 1948), 386.

This article is about those CCC enrolees that lost their lives fighting fires.

287. Rose, Judith. *Timberline Lodge--A Love Story*. Portland, OR: Graphic Arts Center Publishing, 1986.

Timberline Lodge had some of its preliminary work done by the CCC. This book is dedicated to the story of building the lodge.

288. Roth, Bernhard A. "Remember the C.C.C.?" *Soil Conservation* 18 (April 1953), 204-6.

289. Rouse, David S. "Pages From My Past: The Civilian Conservation Corps." *Wisconsin Magazine of History* 71 (Spring 1988), 205-216.

This article is a first-hand account of an enrollee leaving home and going to camp for the first time. He gives a good description of his life experiences first at Camp Devil's Lake and then at Camp Ontario in Wisconsin.

290. Rowland, Howard. "Can the CCC Blaze a New Trail?" *Conservation* 3 (July-August 1937), 24-6.

The author suggests a permanent CCC with certain changes. Some of the changes include enrolling only farm and rural youth rather than urban youth, elimination of the Army and making woodsmanship part of the learning curriculum.

291. Rudeen, Marlys, comp. *The Civilian Conservation Corps Camp Papers: A Guide*. Chicago: Center for Research Libraries, 1991.

This is a compendium of CCC camp newspapers microfiched by the Center for Research Libraries. The volume is divided into three sections consisting of an alphabetical listing of all titles, a list of all companies and their respective camp newspaper titles, and a geographic index. This list is available online. See the "Introduction" section for details.

292. Ryan, J.C. *The CCC and Me*. Duluth, MN: J.C. Ryan, 1987.

293. _____. *How Dark is the Forest?* Duluth, MN: St. Louis County Historical Society, 1983.

294. Salmond, John A. *The Civilian Conservation Corps 1933-1942: A New Deal Case Study*. Durham, NC: Duke University Press, 1967.

295. _____. "Roosevelt's Tree Army: A History of the Civilian Conservation Corps, 1933-1942." Ph.D. diss., Duke University, 1963.

296. Scharff, Virginia J. *Seeing Nature Through Gender*. Lawrence, KS: University Press of Kansas, 2003.

Chapter 5, "New Men in Body and Soul" looks at the development of the male body through the hard work and exercise. It focuses on turning boys into men and recreating manhood in body and soul. Looks at the nation in emerging from the weakness of the Depression into a strong nation.

297. Scheibe, Fritz Hubertus. *The Civilian Conservation Corps: Wesen und Gestalt des Arbeitsdienstes in den Vereingten Staaten von Amerika.* Wurzburg-Aumuhle: K. Triltsch, 1939.

298. Schenck, C.A. "Tells of 'C.C.C.' in Germany." *Forestry News Digest* (June 1934), 7.

299. Schlesinger, Arthur Meier, Sr. *The New Deal in Action 1933-1939*. New York: The Macmillan Company, 1940.

300. Schlesinger, Arthur M., Jr. *The Age of Roosevelt*. (Vol. 2): *The Coming of the New Deal*. Boston: Houghton Mifflin Co., 1959.

301. Schultz, Michael John. *In the Shadow of the Trees*. Thomaston, ME: Dan River Press, 1997.

This is a fictional story about the CCC.

302. "Second Forest Army." *Business Week* (September 16, 1933), 29.

303. "Secretary of Labor's Recommendations on the Problem of Unemployed Youth." *Monthly Labor Review* 40 (June 1935), 1460.

304. "Sedition in the C.C.C." *New Republic* 86 (February 12, 1936), 6.

305. Seligman, Lester G., and Elmer E. Cornwell, Jr., eds. *New Deal Mosaic: Roosevelt Confers With His National Emergency Council*. Eugene, OR: University of Oregon Books, 1965.

306. "Sense and Nonsense." *American Forests* 48 (May 1942), 221-2.

307. Sherraden, Michael Wayne. "The Civilian Conservation Corps: Effectiveness of the Camps." Ph.D. diss., University of Michigan, 1979.

308. _____. "Three Squares A Day: Enrollees in the Civilian Conservation Corps." *Journal of Social Service Research.* 7 (Spring 1984), 21-40.

Sherraden discusses actual conditions experienced by enrollees in CCC camps. His conclusion was both food and living conditions were associated with overall enrollee satisfaction.

309. Showers, E.G. "Graduates of C.C.C." *School Executive* 60 (December 1940), 34-5.

310. Silcox, Ferdinand A. "Country Needs 120 Million Conservationists." *Forestry News Digest* (June 1934), 18. (Excerpts from Radio Address)

311. _____. "Urges Camps to be Fixture in Nation's Life." *Forestry News Digest* (January 1934), 1, 8-9.

312. Simpson, Charles D. and E.R. Jackman. *Blazing Forest Trails*. Caldwell, ID: Caxton Printers, 1967.

313. Smith, Beverly. "Babes in the Woods." *American Magazine* 117 (May 1934), 27-9+.

314. _____. "Go To It, Kid! Show 'Em What You Can Do!" *American Magazine* 116 (July 1933), 58-9+.

315. Smith, C.D. "What Becomes of CCC Men?" *Phi Delta Kappan* 22 (January 1940), 240.

316. Smith, D. "CCC Plays Safe." *Phi Delta Kappan* 24 (March 1942), 253-4+.

317. Smith, L.F. "Day in the Civilian Conservation Corps." *Internationale Zeitschrift fur Erziehung* 6:4-5 (1937), 270-8.

318. Smith, Langdon, Jr. "The Democratization of Nature: State-Park Development During the New Deal." Ph.D. diss. University of Kansas, 2002.

This work deals with the huge task undertaken by the National Park Service "packaging" nature for mass use by the public. It also looks at the concept of having recreational parks where they benefit all citizens, the idea the author calls the "democratization of nature."

319. Smith, M.L. "Evenings at Cuyamaca." *Sierra Education News* 36 (February 1940), 24+.

320. Smith, Richard Gary. "A Brief History of the Civilian Conservation Corps, 1932-1937." Master's thesis, Southwest Texas State University, 1966.

321. Smith, William H. "Camps or Streets." *School and Society* 41 (February 9, 1935), 196-7.

322. Snow, E.A. and J. Roeser, Jr. "National Forests of the Black Hills." *Black Hills Engineer* 26 (December 1940), 245-59.

323. Sokolsky, George E. "About the New Deal." *American Magazine* 117 (June 1934), 53+.

324. Sonderegger, V.H. "The Civilian Conservation Corps." *Louisiana Conservation Review* 4 (January 1934), 49-51.

325. Stann, S.J. "Recreation in the CCC Camps." *Journal of Health and Physical Education* 12 (November 1941), 517+.

326. Starke, Tom. "The Nazi Soldiers of Camp Albemarle." *Magazine of Albemarle County History* 51 (1993), 131-7.

Camp Albemarle, a former CCC camp in Virginia, became a German prisoner of war camp during WW II. This article looks at how the CCC camp was adapted to hold POW's.

327. "The State Governors Reply to Roosevelt." *Forestry News Digest* (June 1933), 4-5.

328. Stave, Bruce M. *The New Deal and the Last Hurrah.* [Pittsburgh, PA]: University of Pittsburgh Press, 1970.

329. Steely, Jim W. "The Architectural Legacy." *Panhandle-Plains Historical Review* 67 (1994), 51-70.

330. Stepenoff, Bonnie. "Archives and Historic Preservation: The Case of the CCC." *Midwestern Archivist* 13:2 (1988), 77-83.

This article discusses the survey of CCC structures in Missouri state parks conducted by the Missouri Historic Preservation program. Also discussed was the role of archivists and historians in historic preservation, the research involved, and a final product in 1985, when Missouri became the first state to produce a thematic nomination of ECW resources to the National Register of Historic Places. A total of 342 buildings were included in 42 districts and 32 individual property listings.

331. Stevens, E.J. "Part-Time School for C.C.C. Boys." *Agricultural Education Magazine* 10 (December 1937), 110.

332. Story, Isabelle F. *Glimpses of Our National Parks*. Washington, DC: Government Printing Office, 1936.

This pamphlet gives a brief history of the National Park Service and some of the activities of the more than 400 CCC camps in the parks. It provides a brief narrative on many of the larger parks and a listing of the National Military Parks. In the back of the pamphlet is a map showing the location of the National Park Service projects.

333. _____. *The National Parks and Emergency Conservation Work*. Washington, DC: Government Printing Office, 1936.

This pamphlet is very similar to the preceding one. It looks at the work of the CCC in national and state parks.

334. Street, Gerald. "Wisdom From the Trees." *American Forests* 40 (January 1934), 24-5.

335. "Strength of C.C.C. Set at 428,000." *American Forests* 42 (January 1936), 35.

This short article provides information regarding the decrease in CCC enrollment beginning January 1, 1936.

336. Stuart, Robert Y. "That 250,000-Man Job." *American Forests* 39 (May 1933), 195.

337. Swain, Donald C. *Wilderness Defender: Horace M. Albright and Conservation*. Chicago: University of Chicago Press, 1970.

Albright was Director of the National Park Service at the beginning of the New Deal. Several mentions are made in this book about the work done by the CCC and their role in conservation.

338. Taylor, Jed H. "After the CCC--What?" *Journal of Adult Education* 9 (June 1937), 303-4.

339. Thorpe, Lloyd. *Men to Match the Mountains*. n.p.: Lloyd Thorpe, 1972.

340. "Three Governors Look to Future of Work Begun by C.C.C." *Forestry News Digest* (June 1934), 3.

341. Throop, Elizabeth Gail. "A Characteristic Expression: A Thematic Evaluation of Forest Service Depression-Era Administrative Buildings in the Pacific Northwest." *Contract Abstracts and CRM Archaeology* 3 (1983), 123-9.

342. Tilden, Freeman. *The State Parks: Their Meaning in American Life*. New York: Alfred A. Knopf, 1962.

343. Tinker, Earl W. "Big Forestry Opportunity." *Forestry News Digest* (August 1933), 1+.

344. _____. "President's Emergency Conservation Work: With Discussion." *Journal of Forestry* 32 (February 1934), 208-18.

345. Titis, Harold. "Will We Make the Same Mistakes Again?" *Saturday Evening Post* 217 (September 9, 1944), 19+.

346. "To Be or Not To Be." *American Forests* 48 (February 1942), 53.

347. Todd, A.J. "Social Implications of the CCC." *Junior-Senior High School Clearing House* 10 (November 1935), 152-8.

348. Toleen, Leslie E. *Memories of Civilian Conservation Corps: 1939-1942*. [Chisholm, MN: Iron Range Research Center], 1992.

349. "Valor Awards." *American Forests* 43 (July 1937), 356.

Lists citations given to enrollees for performing deeds of heroism.

350. "Voluntary Political Contributions to Indiana." *Journal of Forestry* 32 (November 1934), 874-6.

351. Walston, Kathleen Anne. "A Case for Revival of the Civilian Conservation Corps in Florida." Master's thesis, University of Florida, 1986.

352. Warner, Jean Shumway. "A Policy Study of Youth Service: Synthesizing Analysis of Policy Content and Policy Process Over Time." Ph.D. diss., University of Oklahoma, 1995.

353. Wasch, Diane Shaw, Perry Bush, Keith Landreth, and James Glass. *World War II and the U.S. Army Mobilization Program: A History of 700 and 800 Series Cantonment Construction.* Washington, DC: National Park Service, Cultural Resources, HABS/HAER, 1996.

This book discusses housing for military installations, but mentions how the concept of these style barracks were based ont he earlier CCC barracks.

354. Waugh, Frank A. *Landscape Conservation.* Washington, DC: Emergency Conservation Work, 1935.

355. Wecter, Dixon. *The Age of the Great Depression.* New York: The Macmillan Company, 1948.

356. Wharton, George W., Jr. "C.C.C. Camp 334." *American Motorist* 8 (August 1933), 12+.

357. "When Peace Comes." *Journal of Forestry* 39 (April 1941), 347-8.

This article looks at the question of what are we going to do about our forests and the conservation program after the looming war is over.

358. "Who's Who in the Connecticut CCC?" *Connecticut Woodlands* 3 (January 1938), 8-9.

359. "Why Baby the CCC?" *Collier's* 108 (October 11, 1941), 90.

360. Williams, A. "Youth and Our Economic Problem: What are We Going to do About It?" *Vital Speeches* 5 (August 15, 1939), 659-62.

361. Williams, Aubrey. "The Government's Responsibility for Youth." *Annals of the American Academy of Political and Social Science* 194 (November 1937), 119-28.

Williams talks about the vocational education program in the CCC and its relationship to other New Deal education programs, such as the National Youth Administration.

362. Williams, Gerald W. "The Civilian Conservation Corps: A Pragmatic Response to the Great Depression." *The Umpqua Trapper* 19 (Fall 1983), 51-60

363. _____. *Selected References Concerning the USDA Forest Service.* Roseburg, OR: Umpqua National Forest; Eugene, OR: Willamette National Forest, 1985.

364. Williams, R.E. "Conservation of Human Resources in CCC Camps." *Minnesota Journal of Education* 21 (January 1941), 217.

365. Wilson, Charles M. "Troopers of the Fog." *Commonweal* 18 (September 15, 1933), 462-5.

366. Winthrop, Henry. "Problems and Prospects of the Intellectually Underprivileged." *Journal of Human Relations* 505-23.

The author discusses the revival of the CCC because of its success with employing American intellectually underprivileged youth in marginal services of conservation and other outdoor occupations.

367. Winton, G.B. "Army and the CCC: Reply to J.H.H. Darling." *Christian Century* 50 (November 8, 1933), 1411-12.

368. Wirth, Conrad L. *The Civilian Conservation Corps Program of the United States Department of the Interior, March 1933 to June 30, 1943.* Chicago: National Park Service, 1944.

This is the final report of the CCC, by Wirth, to the Secretary of the Interior. It gives a brief description of the tasks and work programs of each individual agency involved in the CCC. It also provides a good history of the make up and administration of the CCC program.

369. _____. *Parks, Politics and the People.* Norman, OK: University of Oklahoma Press, 1980.

370. Wirth, Conrad L. and James F. Kieley. "It's 50 Years Since CCC went into Action." *NPS Courier* (April 1983).

371. Witt, Wm. H. "The Civilian Conservation Corps on a Youth Reconservation Corps." *Military Surgeon* 83 (November 1938), 445-52.

372. "Woodring Sees Million CCC's Next Summer." *U.S. Forest Service Bulletin* 18 (February 12, 1934), 6-7.

373. Woods, James Russell. "The Legend and the Legacy of Franklin D. Roosevelt and the Civilian Conservation Corps (CCC)." Ph.D. diss., Syracuse University, 1964.

Woods discusses Roosevelt's involvement in fostering the CCC. He also includes a chronology of Roosevelt's decisions and concerns for the administration and continuation of the CCC from 1932 through 1944.

374. Yenawine, Wayne S. "Civilian Conservation Corps Camp Papers, A Check List of Papers Published Before July 1, 1937, With Notes." Master's thesis, University of Illinois, 1938.

375. "Youth in Industry," *Monthly Labor Review* 52 (June 1941), 1405-13.

376. Zerbe, Farren II. "I Have Lived With Nature--Man's Best Friend and His Last." *American Forests* 40 (April 1934), 164+.

Zerbe talks about his experiences in the CCC and working with nature.

ADMINISTRATION

377. "Anti-Pneumonia Vaccine Being Distributed to C.C.C. Camps." *Science News Letter* 32 (November 20, 1937), 324.

378. "Army Develops Civilian Supervision of C.C.C. Camps." *American Forests* 40 (May 1934), 224-5.

379. Bassett, Octavia R. "Health and Culture in the CCC Camps." Master's thesis, George Washington University, 1938.

380. Bennett, T.C. "Selective Placement Program." *University of Pennsylvania School of Education Coordination of School and Community,* 205-8.

381. Bethune, Ed. *Characteristics of Civilian Conservation Corps Selectees, October, 1937.* Little Rock, AR: Department of Public Welfare, 1937.

382. "Big C.C.C. Moving and Housing Plan Completed." *Forestry News Digest* (October 1933), 1-3.

383. Bishop, Freeman C. "Strength of Forestry Corps May be Increased to Million." *Evening Star*, Washington, D.C. (November 12, 1933), A-4.

384. Bohn, John J. "Civilian Conservation Corps." *Cavalry Journal* 42 (July-August 1933), 41-2.

This article tells how the War, Interior and Labor Departments mobilized the CCC. It tells how plans were changed to meet the needs and how these needs were greater that the mobilization effort for World War I. Bohn mentions that when different departments of government need to cooperate and work together, they can.

385. Bond, Paul S. "Why Waste the Opportunity?" *Military Engineer* 27 (May-June 1935), 225.

386. Bradley, Robert B. "The Role of the Army in Pensacola." *Pensacola History Illustrated* 2:4 (1987), 22-8.

This article gives a good history of how the peacetime army operated in Pensacola. It further explains how the addition of the CCC to the army's duties increased expenditures and boosted the local economy.

387. Brown, Sidney G. "Army Experience Aids Forest Camps." *Washington Post* (June 11, 1933), 5.

388. Browning Hardy P. "Signal Communications for the Civilian Conservation Corps." *Signal Corps Bulletin* (July-August 1936), 21-24, 26.

389. Buel, Walter S. "Army Under the New Deal: Non-Military Mobilization of the Civilian Conservation Corps." *Literary Digest* 116 (August 26, 1933), 2-4+.

390. "CCC Also Spends: Expansion of Conservation Camps." *Business Week* (May 4, 1935), 12+.

391. "CCC and WPA." *American Forests* 47 (July 1941), 340-1.

392. "C.C.C. Awards of Valor." *American Forests* 42 (July 1936), 335.

This is the first list of valor awards presented to CCC enrollees or former enrollees who saved lives while in the CCC. The name and a brief summary of the heroic deed is included for each of these eleven men.

393. "CCC Camps Reduced." *American Forests* 47 (May 1941), 242.

394. "CCC Dons New Uniforms." *American Forests* 45 (March 1939), 132.

395. "CCC's Doom." *Newsweek* 19 (June 15, 1942), 30.

396. "C.C.C. Enrollment Increasing." *American Forests* 41 (September 1935), 552.

397. "C.C.C. Expansion Gets Under Way." *American Forests* 41 (June 1935), 286.

398. "CCC Extended for Three More Years." *American Forests* 43 (August 1937), 401+.

This is the full text of the provisions of the new act "To Establish a Civilian Conservation Corps, and for Other Purposes" as passed by Congress.

399. "CCC Food Bill is 6 Million A Month." *Forestry News Digest* (January 1936), 20.

400. "C.C.C. Makes Ready for Expansion." *American Forests* 41 (March 1935), 146.

401. "CCC Needs Clearer Policy on Conservation." *American Forests* 44 (May 1938), 224.

402. "C.C.C. Ordered to Prepare for New Period." *American Forests* 41 (April 1935), 178.

403. "CCC Spent $29,375,938 for Trucks, etc. in Six Months." *Automotive Industry* 73 (December 14, 1935), 783.

404. "C.C.C. to be Extended." *American Forests* 40 (February 1934), 81.

405. "CCC to Build 350 Winter Camps Costing $6,500,000--Corps Reaches Peak Enrollment of 359,870." *Commercial and Financial Chronicle* 139 (August 25, 1934), 1184.

406. "C.C.C. Toilet Kit Purchase Approved." *American Forests* 39 (July 1933), 322.

407. "CCC Will be Continued Another Six Months." *Commercial and Financial Chronicle* 138 (February 17, 1934), 1172.

408. "C.C.C. Work Accomplishments for First Year." *American Forests* 40 (July 1934), 322.

409. Carnes, William G. "Landscape Architecture in the National Park Service." *Landscape Architecture* (July 1951), 145-150.

410. "Causes of Death: U.S. Army Compared with CCC." *Army Medical Bulletin* 46 (October 1938), 100-5; vol. 48 (April 1939), 113-18; and, 53 (July 1940), 114-19.

411. Coakler, Robert W., Richard C. Kugler and Paul J. Scheips. *Army Experience in Youth Training, 1920-1955.* Washington, DC: Office of the Chief of Military History, 1964.

412. Cohn, Samuel F. "A CCC Commissary: How Organized and Operated." *Quartermaster Review* 14 (July-August 1934), 46-7, 73-4.

413. "Congress Abolishes CCC." *Commercial and Financial Chronicle* 156 (July 9, 1942), 115.

414. "Congress Asked to Make CCC Permanent." *American Forests* 43 (May 1937), 244.

415. "Conservation Corps Starts its Second Half Year." *News Week* 2 (October 7, 1933), 10.

416. "Conservation in a Post-War World." *American Forests* 47 (October 1941), 469.

Again, this is an article that looks at whether or not the country will go bact to a conservation program after the pending war is over.

417. "Conserving Human Resources." *Survey* 71 (March 1935), 81.

418. Cook, Allen. "Dawn of a New Day." *American Forests* 39 (December 1933), 531-4.

Cook looks at the opportunity of conservation and reforestation and the help that will be provided by the CCC.

419. "Corps Recruits for New Period." *American Forests* 39 (November 1933), 516.

420. Cowan, C.S. "C.C.C. Movement and its Relation to the Clarke-McNary Act." *Journal of Forestry* 34 (April 1936), 383-7.

Cowan suggests that the fire protection funds allocated under the Clarke-McNary Act could be put to better use in a fire protection force and not the CCC. The CCC should not be relied upon to fight all the forest fires. Trained local fire fighters would be a better value for the money spent. He bases his theory on the findings of the Copeland Report.

421. Cox, W.C. "Outbreak of Epidemic Cerebrospinal Meningitis in a CCC Camp." *American Journal of Public Health* 25 (July 1935), 829-32.

422. "Critics Keep Their Hands Off the CCC." *Literary Digest* 118 (August 18, 1934), 8.

423. Crutchfield, Lee G. Jr. "It Has Given Me a Niche to Fill." *American Forests* 40 (February 1934), 66-7.

Crutchfield talks about his experience in the CCC, and the place it has provided him in conserving the forest.

424. Davison, Stanley. "Establishment of the First Year's Administration of the Civilian Conservation Corps." Master's thesis, University of Montana, 1935.

425. Dearborn, Ned H. *Once in a Lifetime: A Guide to the CCC Camp.* New York: Charles E. Merrill Co., 1936.

This is probably one of the best descriptive sources on the CCC in print. The first part, on the CCC in general, is in question and answer style. Then the book lists and explains all of the job/learning possibilities in the CCC.

426. Derderian, William John. "From Idea to Reality: A Study of Operations in the Civilian Conservation Corps, 1933-1942." Master's thesis, California State University, Sacramento, 1962.

427. Dern, George H. "Says C.C.C. Free of 'Militarism' in Praise of Speech Marsh Made." *Forestry News Digest* (April 1934), 1+.

428. Dingley, Nelson. "Let the Army Do It." *Coast Artillery Journal* 76 (July-August 1933), 265+.

429. Dolman, Paul H. "Why Is the CCC?" *Reserve Officer* 12 (October 1935), 17, 19.

430. Donovan, Leo. "The Establishment of the First Conservation Corps." *Infantry Journal* 40 (July-August 1933), 245-49.

431. Dubay, Robert W. The Civilian Conservation Corps: "A Study of Opposition, 1933-1935." *Southern Quarterly* 6 (April 1968), 341-58.

432. Edson, C.A. "What's to Become of the CCC?" *Forum* 93 (April 1935), 245-7.

433. "Eight Years of CCC Operations, 1933-1941." *Monthly Labor Review* 52 (June 1941), 1405-13.

434. "Emergency and Experimental New Deal Legislation Expiring this Year." *Congressional Digest* 16 (January 1937), 2.

435. "Emergency Conservation Program." *Science* 78 (July 21, 1933), 52.

436. "Emergency Conservation Work." *Congressional Record* 78 (Letter Addressed to the President, December 22, 1933, by Robert Fechner, Director). 177-80.

437. "Emergency Conservation Work." *Monthly Labor Review* 39 (September 1935), 780-1, (November 1935), 1270; 40 (January 1935), 212-4, (February 1935), 467-8, (March 1935), 798+.

438. "Emergency Conservation Work During 1933." *Monthly Labor Review* 38 (March 1934), 518-22.

439. "End of CCC." *Time* 39 (June 15, 1942), 10.

440. "Enlargement of C.C.C. Planned." *American Forests* 41 (February 1935), 91.

441. "Enlisting the Jobless to Smash the Depression." *Literary Digest* 115 (April 1, 1933), 5.

442. "Enroll 50,000 in C.C.C." *American Forests* 42 (August 1936), 380.

443. "Enrollment in the Civilian Conservation Corps." *Congressional Record* 79 (August 22, 1935), 14522.

444. "Enrollment Lagging." *American Forests* 41 (August 1935), 388.

445. Falk, Stanley. Untitled Research Document Pertaining to the Army's Experience with the Civilian Conservation Corps. Washington, DC: U.S. Center for Military History, Historical Research Collection, 1964.

446. "Fechner Announces New C.C.C. Program." *American Forests* 42 (November 1936), 524.

This announcement provides for an extra 16,200 new enrollees from the drought states in the west and south, but still maintaining total strength of the CCC at 350,000.

447. "Fechner Announces Reduction of CCC Camps." *American Forests* 44 (April 1938), 172.

448. "Fechner Clarifies Civilian Status of CCC." *American Forests* 45 (October 1939), 511.

449. "Fechner Inspecting Western C.C.C. Camps." *American Forests* 40 (August 1934), 372.

450. "Fechner Recommends Permanent C.C.C. in Report." *American Forests* 41 (May 1935), 240.

451. "Fechner Report Recommends Permanent C.C.C." *American Forests* 41 (May 1935), 240.

452. "Fechner Reviews Six Years of CCC Conservation Work." *American Forests* 45 (August 1939), 418.

453. Fechner, Robert. "Believes C.C.C. to Continue." *Evening Star*, Washington, D.C. (October 25, 1934), A-10.

454. _____. "The Civilian Conservation Corps of the U.S.A. (Working Camps for Unemployed Youth)." *International Observer* 1 (1937), 132-44.

455. _____. "Civilian Conservation Corps Program." *Annals of the American Academy of Political and Social Science* 194 (November 1937), 129-40.

456. _____. "Expansion of CCC A Big Undertaking." *New York Times* (June 16, 1935), 10-E.

457. _____. "Food Inspection for the Civilian Conservation Corps." *Journal of the American Veterinary Medicine Association* 21 (September 1937), 276-87.

458. _____. "A Mighty Force in the CCC." *Army Chaplain* (January-February, 1939), 75-6.

459. _____. "Safety in the Three C's." *Safety Engineering* 76 (October 1938), 19-20.

460. "Fechner to Recommend Continuation of C.C.C." *American Forests* 40 (October 1934), 490.

461. "Fechner Urges Greater C.C.C." *American Forests* 41 (January 1935), 44.

462. "Federal Act Providing for Relief of Unemployment Through Reforestation." *Monthly Labor Review* 36 (May 1933), 1039-40.

463. "Feeding the CCC: A Big Market for Food Manufacturers." *Food Industries* 8 (February 1936), 59.

464. "58th Annual Meeting Sounds 'New Deal' for Forestry." *American Forests* 39 (October 1933), 456-7+.

465. "First of Army of Idle Now in Camp." *News Week* 1 (April 15, 1933), 6.

466. FitzGerald, Francis V. "The President Prescribes: Organization and Supply of CCC." *Quartermaster Review* 13 (July-August 1933), 7.

467. Fitzpatrick, J.L. "Conserving Forests and Building Men: CCC as Part of the New Deal." *Scholastic Magazine* 24 (March 17, 1934), 23-5.

468. "Forest and Camp: Congressional Controversy Over Militarism." *Literary Digest* 123 (January 16, 1937), 10.

469. "Foresters Urge Change in Handling CCC." *Forestry News Digest* (January 1936), 3.

470. "The Forestry Corps on Review." *American Forests* 39 (November 1933), 497.

471. Foss, Elmer T. "Eyes on the Ball...How The Quartermaster General Keeps Track of the CCC Camps." *Quartermaster Review* 13 (September-October 1933), 35.

472. "Future of CCC." *American Forests* 47 (March 1941), 130+.

473. "Future of CCC and NYA." *Nature Magazine* 35 (January 1942), 33-4.

474. Gilbertson, G. H. "Rookie, Watch Your Step: Safety in the CCC." *Safety Engineering* 83 (February 1942), 12-4.

475. Golze, Alfred R. "Emergency Conservation Work Safety Program." *Reclamation Era* 26 (June 1936), 175.

476. _____. "The Future of the Civilian Conservation Corps." *Reclamation Era* 27 (November 1937), 266-7.

477. Gorham, Eric. "The Ambiguous Practices of the Civilian Conservation Corps." *Social History* 17 (May 1992), 229-49.

Gorham suggests that the makeup of the CCC was state-controlled autonomous camps where each aspect of the enrollees life was predetermined. He further states that these camps kept enrolles in a controlled environment where each individual knew his place and what was expected of him. This article explores the concept that the CCC was different from all other New Deal programs because it only focused on young, mainly rural, out of work males. Very few minorities were included, and all "exemplified the normalizing potential of a democratic state."

478. Graf, Enoch. "The Army's Greatest Peace-Time Achievement." *Quartermaster Review* 15 (July-August 1935), 1-13.

This article looks at the Army's administration of the CCC. It shows the logistics of supplying and administering a company of enrollees.

479. Greene, Paul C. "Suitable for Duty with Civilian Components. Why Does Regular Shy Away from CCC Duty?" *Infantry Journal* 44 (March-April 1937), 141-42.

480. Griffith, Robert K., Jr. "Quality not Quantity: The Volunteer Army During the Depression." *Military Affairs* 43:4 (1979), 171-7.

481. Guthrie, John D. "Civilian Conservation Corps." *Military Engineer* 26 (January-February 1934), 15-19.

Guthrie, a major in the U.S. Army Corps of Engineers, provides a look at the organization and accomplishments of the CCC during its first three months of operation in 1933.

482. Hagood, Johnson, Jr. *Company Commanders' Handbook for the C.C.C.* Fort Sam Houston, TX: n.p., 1935.

This book lists all of the rules and regulations for administering a CCC company. It includes all forms and examples of how to fill them out.

483. _____. "Soldiers of the Shield: The Civilian Conservation Corps Stands Square with National Defense." *American Forests* 40 (March 1940), 103-5.

Hagood explains how West Point officers, Citizens' Military Training Corps and Officers' Reserve Corps work. He also tells how each fits into the CCC administration.

484. Hammons, Vernon L. *A Brief Organizational History of the Office of Design and Construction, National Park Service, 1917-1962.* [Washington, DC: National Park Service, n.d.]

This history, although not specifically CCC, tells who was who in the Design and Construction Division of the National Park Service. It is an administrative history which includes many people who were prominent in the rustic style of architecture used in national parks.

485. Harper, Charles Price. *The Administration of the Civilian Conservation Corps.* Clarksburg, WV: Clarksburg Publishing Co., 1939.

486. _____. "The Administration of the Civilian Conservation Corps." Ph.D. diss., Johns Hopkins University, 1937.

487. Harrison, Laura Soulliere. *Architecture in the Parks: National Historic Landmark Theme Study.* Washington, DC: National Park Service, 1986.

As the title indicates this is a National Historic Landmark study of rustic architecture and use of native materials in the construction of national park service buildings and sites. Many National Register Nomination forms are included and contain history, pictures and maps of the national parks included in the study.

488. _____. *Historic Housing in the National Park System.* Washington, DC: National Park Service, 1990.

This very informative study was performed as a result of Special Directive 88-3: Employee Housing–Responsibilities of Management. Housing constructed by the CCC is only part of the 1,082 structure report.

489. Haskin, Frederick J. "The Army's New Mobilization Record." *Evening Star*, Washington, D.C. (August 27, 1933), 2.

490. Hawes, Austin F. "Plan to Establish CCC Graduates in Rural Communities." *Connecticut Woodlands* 6 (January 1941), 8-9.

491. _____. "Readjusting the C.C.C. as a Permanent Social Factor." *Journal of Forestry* 32:9 (December 1934), 937-41

Hawes expresses his concerns about selecting high quality men for supervisory positions with the CCC. He also looks at the social impacts of those men selected and their qualifications.

492. Hay, Howard. "Military Training for CCC." *American Forests* 45 (May 1939), 242.

493. Headington, L.H. "A Study of the Duties of Company Commanders, Educational Advisers, and Project Superintendents in the Civilian Conservation Corps of West Virginia." Master's thesis, Ohio State University, 1938.

This study, though undertaken in West Virginia, could have been done anywhere in the country. It is enhanced with graphs and charts and discusses the duties of those in charge of the camps in detail

494. "Health in the CCC Camps." *Science* 84 (July 10, 1936), sup. 7.

495. Henry, Lyle K. "Civilian Conservation Corps as an Educational Institution." *School and Society* 43 (January 11, 1936), 62-6.

496. _____. "Evaluation of the Camp Program and its Relationship to Public Education." *Internationale Zeitschrift fur Erziehung* 6 (1937), 278-9.

497. Hill, Frank E. "Salvaging Youth in Distress." *New York Times Magazine* (April 21, 1935), 12-13.

498. Hiller, F.H. "Probationers in the C.C.C. Camps." *National Probation Association Yearbook* (1935), 121-30.

499. Hitchens, A.P. "Experience of the Army and Civilian Conservation Corps in Handling Newly Mobilized Men." *American Journal of Public Health* 30 (November 1940), 1297-1301.

500. Holbrook, Stewart. "Forest Architecture--Old and New." *American Forests* 42 (December 1936), 558-61.

This pictorial gives a short history of forest buildings, from the early ranger stations, housing and park structures through the newer ones built by the CCC.

501. "House Extends CCC as Temporary Agency for Period of Two Years." *Commercial and Financial Chronicle* 144 (May 15, 1937), 3260.

502. Hoyt, Ray. *"Your CCC:" A Handbook for Enrollees.* Washington, DC: Happy Days Publishing Company, 1938.

503. Hull, James Howard. "Our Permanent Conservation Corps: An Officer Formerly Attached to the C.C.C. Gives A View of the Camps Which Sunday Visitors Don't See and Explains Why it Costs More than it's Worth." *Nation's Business* 28 (May 1940), 23-4+.

504. Hurd, Charles W.B. "Forestry Army at the Front." *Literary Digest* 116 (September 9, 1933), 5-6.

505. Irwin, William Grover. "The Work of the Chaplain in the Civilian Conservation Corps." B.D. thesis, Emory University, 1940.

506. Jernberg, Albert W. *My Brush Monkeys: An Army Officer's Story of the C.C.C.* New York: Richard R. Smith, 1941.

507. Johnson, Charles W. "The Army and the Civilian Conservation Corps, 1933-1942." *Prologue, The Journal of the National Archives* 4 (Fall 1972), 139-156.

Johnson discusses the strain on the army in the early days of the CCC. He says the requirements for regular army officers to command CCC units were over 50% of those serving on active duty at the time. He continues by saying the problem was lessened by the substitution of regular officers by reserve officers and the lessons learned by mass mobilization. Also discussed is the strain on national defense, service school personnel and the lack of military discipline over civilian enrollees.

508. Johnson, Charles William. "The Civilian Conservation Corps: The Role of the Army." Ph.D. diss., University of Michigan, 1968.

509. Johnson, Jesse J. *Court-Martial in the Shadow of the Trees: Blacks in the Civilian Conservation Corps, 1934; Time 1934, Location Pennsylvania.* Hampton, VA: J.J. Johnson, 1994.

510. Judson, Lucius V. *Development of the Civilian Conservation Corps Program with a Personal Analysis of One Hundred Fifty Enrollees.* Ann Arbor, MI: n.p., 1942.

511. Kaiser, Clifford A. *Group Feeding.* n.p., n.p., 1939.

512. Kililae, Walter. "CCC Training for Reserve Officers." *Cavalry Journal* 49 (September-October 1940), 456-7.

513. Killigrew, John W. *The Impact of the Great Depression on the Army.* New York: Garland Publishing, 1979.

514. _____. "The Impact of the Great Depression on the Army, 1929-1936." Ph.D. diss., Indiana University, 1960.

Killigrew's dissertation looks at the problems associated with the army's administration of the CCC. He covers the problems associated with the Great Depression and the call up of reserve officers to administer the CCC camps. Discussed also is the need for the army's corps area control of supply and administration to facilitate training and staffing.

515. Kirkpatrick, C.W. "No Militarism in the CCC." *Christian Century* 52 (April 24, 1935), 549.

516. Knudsen, C. "Different Reports on the C.C.C.: Reply to L.E. Tharpe." *Christian Century* 50 (September 13, 1933), 1149.

517. Kuhns, D.M. "Health in the CCC Camps: Control Measures Against Meningitis Epidemic." *Science* 84 (July 10, 1936), Sup. 7-8.

518. Lull, George F. "Fevers of the Typhoid Group in Members of the Civilian Conservation Corps During 1934." *American Journal of Public Health* 25 (July 1935), 839-40

519. _____. "Typhoid in the CCC." *Military Surgeon* 79 (July 1936), 45-48.

520. Major, Duncan K. "Mobilizing the Conservation Corps." *Army Ordnance* 14 (July-August 1933), 33+.

521. "Making the C.C.C. Permanent." *American Forests* 40 (July 1934), 311.

522. "Marsh Resigns as C.C.C. Educational Director." *American Forests* 41 (March 1935), 166.

523. McConkey, Darel. "The Health of John Peavey." *American Forests* 40 (November 1934), 509-11+.

This article looks at the healthful work that is being done by the enrollees working in the forests.

524. "McEntee Named Head of CCC." *American Forests* 46 (March 1940), 129.

525. "Medals Given to C.C.C. Officers." *American Forests* 40 (January 1934), 38.

This short article tells of the camp commanders, one each from each of the nine corps areas, who were selected as having the best camp in the corps area. It lists each officer, and tells that they went to Washington, DC, to receive their awards.

526. "Meeting a Post-War Need." *American Forests* 47 (November 1941), 527.

This article discusses a post-war need for similar conservation programs.

527. "Men and Mountains Part." *American Forests* 48 (August 1942), 363.

This article talks about the enrollees that are preparing to go to war and leave the forests for the army.

528. Mergen, Bernard. "Denver and the War on Unemployment." *Colorado Magazine* 47:4 (1970), 326-337.

529. "Militaristic C.C.C." *World Tomorrow* 17 (January 18, 1934), 29-30.

530. "More CCC Camps Close." *American Forests* 47 (November 1941), 534.

531. Morrell, Fred. "Emergency Conservation Work Policies." *Journal of Forestry* 34 (March 1936), 284-92.

This article uses charts to tell the distribution of enrollees by age, location and work project. It also provides statistics on federal, state and provately owned land work projects, by region.

532. "Morrell on CCC Advisory Council." *American Forests* 44 (May 1938), 226.

533. "Move to Consolidate CCC and NYA." *American Forests* 47 (December 1941), 589.

534. "National Park Standards: A Declaration of Policy." *National Parks Bulletin* 14 (December 1938), 5-7.

535. "New Civil Service Jobs for C.C.C." *American Forests* 42 (December 1936). 574.

This announcement tells of a new civil service position, that of Junior Assistant to Technician, being added to the civil service list. Approximately 15,000 enrollees passed the exam and became eligible for the new position that paid $85 per month.

536. "New Forest Fire Safety Regulations for CCC." *American Forests* 45 (October 1939), 516.

537. "Officials Blamed in Death of CCC Fire Fighters." *American Forests* 44 (December 1938), 566.

538. "Operations in 1936." *Monthly Labor Review* 44 (April 1937), 896-8.

539. Oxley, Howard W. "Analysis of Enrollee Personnel." *School Life* 26 (March 1941), 187-8.

540. _____. "Basis of Curriculum Planning." *School Life* 25 (April 1940), 220-1.

541. _____. "Building A Model Camp Program." *School Life* 22 (November 1936), 80-1.

542. _____. "Camp Life as a Basis for Program Planning." *School Life* 26 (November 1940), 56-7.

543. _____. "CCC Camps Train Young Farmers." *School Life* 22 (April 1937), 235+.

544. _____. "Certification of CCC Educational Work." *School Life* 25 (July 1940), 317.

545. _____. "Job of the Camp Adviser." *School Life* 23 (September 1937), 24-5.

546. _____. "Philosophy of CCC Camp Education." *School Life* 24 (March 1939), 181+.

547. _____. "Reviewing the Educational Program." *School Life* 26 (April 1941), 218-9+.

548. _____. "State Education Departments Aid CCC." *School Life* 24 (November 1937), 73+.

549. _____. "Teaching Techniques in the C.C.C." *School Life* 21 (January 1936), 112.

550. _____. "Training of Camp Educational Advisers." *School Life* 24 (February 1939), 137+.

551. _____. "Training the Camp Adviser." *School Life* 21 (April 1936), 208+.

552. _____. A University of the Woods." *Conservation* 1 (October 1935), 8.

Tells of the educational opportunities available to CCC enrollees, whether it be handicrafts, high school or college courses or social or economic skills. The article also tells of the variety of needs of the individual enrollee.

553. Pack, C.L. "Auditing the C.C.C. Ledger." *Review of Reviews and World's Work* 89 (January 1934), 28-9+.

554. Paige, John C. *The Civilian Conservation Corps and the National Park Service, 1933-1942, An Administrative History*. Denver: National Park Service, 1985.

555. Parker, Homer C. "Conduct of the C.C.C. Camps." *Congressional Record* 78 (January 15, 1934), 668-9.

556. Perley, Reuben Noel. *Today You are a Man.* Omaha, NE: [Reuben Noel Perley], 1940.

This short book is an actual letter written by an army officer to his son on his 21st birthday. It is dedicated to the memory of Robert Fechner, the first director of the CCC. It has a forward written by J.J. McEntee, the current director of the CCC, telling how the CCC is a builder of men.

557. "Permanent CCC Voted by Senate." *Commercial and Financial Chronicle* 144 (May 22, 1937), 3424.

558. Persons, W.F. "Selecting 1,800,000 Young Men for the C.C.C." *Monthly Labor Review* 46 (April 1938), 846-51.

559. Persons, W. Frank. "A Selecting Agency Views C.C.C. Work Opportunities." *Soil Conservation* 3 (March 1938), 240-1.

560. "Plans Shaping to Continue Work Through Winter." *American Forests* 39 (September 1933), 420.

561. "Political Activities in the Civilian Conservation Corps: Committee Report to New England Section, Society of American Foresters." *Journal of Forestry* 31 (December 1933), 914-9; 32 (April 1934), 397-9.

562. Poyet, E. "Hospitalization of the Civilian Conservation Corps." *American Journal of Nursing* 34 (January 1934), 29-30.

563. "President Announces New Program." *American Forests* 41 (October 1935), 600.

564. "President Approves Educational Program for C.C.C." *American Forests* 40 (January 1934), 35.

565. "President Asks Congress to Make CCC Permanent." *Scholastic* 30 (April 24, 1937), 12.

566. "President Backs NYA and CCC Continuance." *Commercial and Financial Chronicle* 155 (March 26, 1942), 1261.

567. "President Considering Educational Program for CCC." *American Forests* 39 (December 1933), 561.

568. "President Makes $20,000,000 Available for Forest Acquisition." *American Forests* 39 (July 1933), 322.

569. "President Orders Continuance of C.C.C. at Strength of 350,000." *American Forests* 42 (May 1936), 234.

States that the number of CCC camps will remain the same but the company strength will be reduced from 200 enrollees to 163. The target goal of 350,000 enrollees will be accomplished by March 31, 1937.

570. "President Orders the Civilian Conservation Corps Doubled." *American Forests* 41 (May 1935), 232.

571. "President Roosevelt Asks Congress to Make CCC Permanent Government Agency." *Commercial and Financial Chronicle* 144 (April 10, 1937), 2462.

572. "President Roosevelt, Prior to Election Indicated Plans to Continue CCC at Minimum Strength." *Commercial and Financial Chronicle* 143 (November 7, 1936), 2932.

573. "President Roosevelt to Request Continuation of Civilian Conservation Corps: Fechner's Report Shows Forestry Program Greatly Advanced by Work." *American Forests* 40 (November 1934), 540.

574. "Principles and Procedures for Selection of Men." *Monthly Labor Review* 40 (May 1935), 1162-9.

575. "Proposal to Combine the CCC and the NYA." *School and Society* 53 (January 4, 1941), 14-15.

576. "Public Health with the Civilian Conservation Corps Work Camps." *American Journal of Public Health* 24 (May 1934), 560-1.

577. "Question of a Permanent C.C.C." *Journal of Forestry* 34 (August 1936), 739-41.

578. "Reclamation Camps Lead CCC in Safety in 1937." *Reclamation Era* 38 (September 1938), 192.

579. "Religion in Conservation Camps." *Missionary Review of the World* 56 (July 1933), 350.

580. "Replacement Enrollment Begins in C.C.C." *American Forests* 41 (November 1935), 652.

581. Reynolds, Harris A. "Need of Junior C.C.C. Camps." *Forestry News Digest* (June 1934), 19.

582. Richardson, Elmo R. "Was There Politics in the Civilian Conservation Corps?" *Forest History* 16 (July 1972), 12-21.

583. Richardson, M.B. "CCC Duty is Engineering Experience." *Mechanical Engineering* 60 (April 1938), 342-3.

584. Ringland, Arthur C. "The Civilian Conservation Corps and the Natural Resources Board." *Journal of Forestry* 32:9 (December 1934), 949-51.

585. Rodriguez, Alexander Frank. "Civilian Conservation Corps: Concept To Mobilization." Master's thesis, San Jose State University, 1993.

Rodriguez tells of the early organizational days of the CCC.

586. Roosevelt, Franklin D. "President's Message." *Louisiana Conservation Review* 3 (April 1933), 2.

587. "Roosevelt Orders 275,000 Men to Conservation Work Camps by July 1." *American Forests* 39 (June 1933), 272-3.

588. "Roosevelt Recommends Permanent C.C.C." *American Forests* 43 (February 1937), 87.

589. "Rule of the CCC: Bars Young Men on Probation and Parole." *Survey* 71 (August 1935), 240.

590. Saalberg, John J. "Roosevelt, Fechner, and the CCC---A Study in Executive Leadership." Ph.D. diss., Cornell University, 1962.

591. "Safety Record Set by C.C.C. Camp." *National Safety News* 39 (June 1939), 70.

592. "Selection of Men for the Civilian Conservation Corps." *Monthly Labor Review* 40 (May 1935), 1162-9.

593. Sellers, S. "CCC Placement Program in the Sixth Corps Area." *School Life* 27 (December 1941), 93-4.

594. Severson, Robert F., Jr. "The Civilian Conservation Corps: A Work Force Solution." *Research in Economic History* Supplement no. 2, 1982, 121-6.

595. Shepard, Nelson M. "Army's Theoretical Preparedness Plan is Successful in Recruiting and Transporting 300,000 Forest Army." *Sunday Star*, Washington, D.C. (June 25, 1933), 1.

596. Sherman, E.A. "Conservation of Our Forests Gains Rapidly on Two Fronts." *New York Times* (December 24, 1933), sec. 8, 6.

597. Sherraden, Michael W. "Military Participation in a Youth Employment Program: The Civilian Conservation Corps." *Armed Forces and Society* 7 (Winter 1981), 227-45.

Sherraden talks of the "unique military management of a public unemployment program in the United States." He states seven criteria in accessing the Army's performance in managing CCC camps and explains his evaluation.

598. Smith, Herbert A. "The Question of a Permanent C.C.C." *Journal of Forestry* 34 (August 1936), 739-41.

599. Smith, Kenneth G. "At A Civilian Conservation Corps Camp." *Recreation* 23 (August 1934), 248+.

600. Smith, Terence J. "The Army's Role in the Success of the CCC." *The Retired Officer* (July 1983), 30-4.

601. Smith, Tom W. "The Polls--A Report: National Service." *Public Opinion Quarterly* 54:2 (1990), 273-85.

602. Stewart, Donald Dean. "The Civilian Conservation Corps: A Study in Institutional Efficiency." Master's thesis, University of Washington, 1940.

603. Swain, Donald C. "The Bureau of Reclamation and the New Deal, 1933-1940." *Pacific Northwest Quarterly* 61 (July 1970).

604. _____. "Harold Ickes, Horace Albright and the Hundred Days: A Study in Conservation Administration." *Pacific Historical Review* 34 (November 1965), 455-65.

Horace Albright was Director of the National Park Service during the first hundred days of the New Deal. This article tells how he shaped conservation policy for the New Deal.

605. _____. "The National Park Service and the New Deal, 1933-1940." *Pacific Historical Review* 41 (August 1972), 312-32.

606. Swing, Raymond Gram. "Take the Army Out of the CCC." *Nation* 141 (October 23, 1935), 459-60.

This article presents the view that when the CCC is made a permanent agency the army control of the camps should be turned over to the U.S. Forest Service. The CCC functioned well under the Army but the Forest Service should supervise conservation efforts.

607. Tanner, P.F. "Youth in the C.C.C." *Catholic Education Review* 40 (September 1942), 406-11.

608. Taylor, Charles H. "Choosing Boys for the CCC." *National Probation Association Yearbook* (1937), 325-40.

609. "10,000 Replacements Ordered for Conservation Corps." *American Forests* 39 (August 1933), 370.

610. Tennessee. Department of Institutions and Public Welfare. *Civilian Conservation Corps: Procedures on Selection of Men.* [Nashville, TN: Tennessee Dept. of Institutions and Public Welfare], 1937.

611. Teston, K. Edd. "A Home and Harvest Saved." *American Forests* 40 (January 1934), 25.

612. Tharpe, L.E. "Conservation Camps and the War Department." *Christian Century* 50 (August 30, 1933), 1089.

613. Theis, H. Jordan. "Yet Another Treasure and Problem for the Army." *Cavalry Journal* 45 (July-August 1936), 292-95.

614. Thiesen, John D. "Civilian Public Service: Two Case Studies." *Mennonite Life* 45:2 (1990), 4-12.

615. "1330 Forest Camps up in Record Time." *Forestry News Digest* (July 1933), 1-2.

616. Thomas, E.D. "CCC: Remarks on Bill to Make the CCC a Permanent Institution." *Journal of the National Education Association* 28 (April 1939), 112.

617. Thone, Frank. "Are We Creating an American Sahara?" *Sunday Star*, Washington, D.C. (August 12, 1934), magazine 6.

618. Tinker, Earl W. "The Role of the United States Forest Service in the Government Emergency Conservation Program." *Central States Forestry Congress. Proceedings* (1933), 108-12.

619. Toner, Raymond J. "Impressions of Duty with the CCC." *U.S. Naval Institute Proceedings* 62 (May 1936), 670+.

620. Tugwell, Rexford G. "New Forest Chief Faced by Difficult Problems." *Sunday Star*, Washington, D.C. (November 19, 1933), pt. 2, 3.

621. Tweed, William C., Laura E. Soulliere and Henry G. Law. *National Park Service Rustic Architecture: 1916-1942.* San Francisco: National Park Service, 1977.

This book discusses the different influences on Park Service architecture up to World War II. Two chapters cover the CCC era, the first telling of the expansion of the park service into various new areas and the decline of manpower in the years leading up to World War II.

622. "Two Years of CCC Work." *Monthly Labor Review* 41 (July 1935), 53-6.

623. Unrau, Harlan D. and Frank G. Williss. *Administrative History: Expansion of the National Park Service in the 1930s.* Denver: Government Printing Office, 1982.

624. Vint T. C. *Report on the Building Program From Allotments of the Public Works Administration, 1933-1937, Eastern Division, National Park Service.* [c. 1938].

625. _____. *Report on the Building Program from Allotments of the Public Works Administration, 1933-1937, Western Division, National Park Service.* [c. 1938].

626. Virginia. State Department of Public Welfare. *A Handbook for Local Selection Agents on CCC Camp Life.* Richmond, VA: Office of Bureau of CCC Selection, Department of Public Welfare, 1940.

627. Vitas, George. "Make the Civilian Conservation Corps Permanent." *Michigan Forester* (University of Michigan Forestry Club) 16 (June 1935), 18-21, 50-4.

628. Wallace, Henry A. "Forestry Faces Challenge says Wallace in his Annual Report." *Forestry News Digest* (December 1933), 1+.

629. ______. "Secretary Wallace, After Survey, Praises Forestry Service." *Washington Post* (August 18, 1935), pt 3, 9.

630. _____. "Tells of Grazing Problems After Trip: Has Good Word for CCC." *Forestry News Digest* (September 1935), 1+.

631. Warren, L.P.D. *Manual of Administration of the Civilian Conservation Corps: For the Use of Company Officers.* Harrisburg, PA: Military Service Publishing Co., 1935.

This was a necessity for the officer overhead in each CCC company. It explained everything from filling out forms to how much food to order for the company. It covered every aspect of company administration for a civilian CCC, since the military was only responsible for operating the camps.

632. Washington (State), Department of Public Welfare. *Civilian Conservation Corps Junior Enrollees: April 1933-July 1935, State of Washington.* Olympia, WA: State of Washington Department of Public Welfare (Monograph No. 6), 1935.

633. Wharton, George W., Jr. "C.C.C. Camp 334." *American Motorist* 8 (August 1933), 12+.

634. X, Captain. "Problems With the Woodland Army." *Military Engineer* 26 (July-August 1934), 297-301.

This article, written early in the CCC program with a pseudonym author, tells of the early experiences of problems encountered by camp commanders and executive officers while establishing their camps in specific locations. The author tells of the great value of independent command and the experience gained in the field, by improvising and solving problems as they arose.

HISTORY

General

635. Albright, Horace M. *Origins of National Park Service Administration of Historic Sites*. Philadelphia: Eastern National Park and Monument Association, 1971.

636. Butler, Ovid. "Prairie Shelter Belt: Thirty Questions and Their Answers Concerning President Roosevelt's Latest Undertaking in Forestry." *American Forests* 40 (September 1934), 395-8.

637. _____. *Youth Rebuilds: Stories from the C.C.C.* Washington, DC: American Forestry Association, 1934.

638. Carpenter, Howard S. *The Civilian Conservation Corps, 1933-1942: A Review*. Washington, DC: U.S. Department of Labor, 1961, (Mimeo-graphed).

639. Everhart, William C. *The National Park Service*. New York: Praeger Publishers, 1972.

640. Evison, Herbert and Newton S. Drury. Interview by Amelia R, Fry. *The National Park Service and Civilian Conservation Corps*. Berkeley, CA: University of California-Berkeley, Regional Cultural History Project, 1963.

641. Fechner, Robert. "The Civilian Conservation Corps Program. *Annals of the American Academy of Political and Social Science* 194 (November 1937) 129-40.

Fechner provides an overall history of the CCC. He also discusses the national distribution of the enrollees and their work projects.

642. Fritz, David L. *Special History Study on Logging and Lumbering as Associated with the Area Now Incorporated Within the Present Bounds of Voyageurs National Park.* Denver: National Park Service, Denver Service Center, 1986.

643. House, Darrell Clore. "The Effect of a Civilian Conservation Corps Camp's Recreational Program on the Realization of its Educational Objectives." Master's thesis, University of Kentucky, 1937.

644. Ickes, Harold L. *Back to Work: The Story of PWA.* Washington, DC: Government Printing Office, 1937.

645. Jesperson, Kathy. "If the CCC Built It, It's Probably Still There." *On Tap* 1 (Fall 2001), 26-30.

This article discusses the CCC in general, but includes material from interviews from some former Pennsylvania enrollees. It lists some Bureau of Reclamation projects in the west, as well as some local projects.

646. Kiefer, E. Kay. *Hobnail Boots and Khaki Suits A Brief Look at the Great Depression and the Civilian Conservation Corps as Seen Through the Eyes of Those who were There.* Chicago: Adams Press, 1983.

647. Kieley, James F., comp. *A Brief History of the National Park Service.* Washington, DC: Civilian Conservation Corps, 1940.

648. Otis, Alison T., William D. Honey, Thomas C. Hogg and Kimberly K. Lakin. *The Forest Service and the Civilian Conservation Corps: 1933-1942.* Washington, DC: USDA Forest Service, 1986.

649. Pandiani, John A. "The Crime Control Corps: An Invisible New Deal Program." *British Journal of Sociology* 33 (September 1982), 348-58.

650. Pfaff, Christine. *The Bureau of Reclamation and the Civilian Conservation Corps, 1933-1942.* Denver, CO: Bureau of Reclamation, 2000.

651. Rauch, Basil. *The History of the New Deal.* New York: Creative Age Press, Inc., 1944.

This book briefly mentions the CCC and its relationship to the New Deal.

652. Rich, Stephen O'Neal. "The Civilian Conservation Corps: A Life Changing Experience." Master's thesis, California State University, Long Beach, 2003.

653. Rutledge, Archibald. "Invading the Sanctuaries." *Conservation* 2 (April 1936), 2.

The author is critical of the CCC for building roads through forests which threaten wildlife. He claims that roads attract visitors, who start fires and force wildlife to leave the area. He thinks other projects, such as dam building, are more important

654. Rutledge, R.H. "C.C.C. Work on Grazing District Lands." *National Wool Grower* 32 (October 1942), 13-5.

655. Stepenoff, Bonnie. "Archives and Historic Preservation: The Case of the CCC." *Midwestern Archivist* 13:2 (1988), 77-83.

656. Stieglitz, Olaf. *100 Percent American Boys: Disziplinierungsdiskurse und Ideologie im Civilian Conservation Corps.* Stuttgart: F. Steiner, 1999.

657. Wilkinson, S. Kristina and Michael R. Hardwick. "La Purisima, A Living Museum of Spanish Colonial History." *Noticias* 9:2 (1973), 1-9.

La Purisima, an old Spanish mission, was rebuilt by CCC enrollees, using the same construction methods as when it was originally built.

658. Wilson, J. "Community, Civility, and Citizenship: Theatre and Indoctrination in the Civilian Conservation Corps of the 1930s." *Theatre History Studies* 23 (2003), 77-94.

659. Winthrop, Kathryn. *Klamath National Forest and the C.C.C.* n.p.: n.p., 1983.

1st Corps Area

Note: Includes Maine, New Hampshire, Vermont, Massachusetts, Rhode Island and Connecticut.

660. Allen, E. John B. "The Development of New Hampshire Skiing: 1870's-1940." *Historical New Hampshire* 36:1 (1981), 1-37.

This article looks at the winter recreation projects developed by the CCC in New Hampshire, especially skiing.

661. Avery, Myron H. "Katahdin: Its History." *In The Maine Woods*. Publication of the Bangor and Aroostook Railroad Company. 1939, 18-26.

662. Berg, Shary Page. *The Civilian Conservation Corps: Shaping the Forests and Parks of Massachusetts, A Statewide Survey of Civilian Conservation Resources* MA: Office of Historic Resources, 1999.

663. Caldwell, John Carroll. "A Curriculum Proposal for a Civilian Conservation Corps Camp in Rural Maine." University of Maine, 1937.

664. Cline, H.C., P.W. Hyres, H.H. Chapman, and H.H. Reynolds. "Political Activities in the Civilian Conservation Corps, Committee Report to New England Section, Society of American Foresters." *Journal of Forestry* 31 (December 1933), 914-9.

665. Draves, David D. *Builder of Men: Life in the CCC Camps of New Hampshire*. Portsmouth, NH: Peter E. Randall, 1992.

666. _____. *The Civilian Conservation Corps in New Hampshire: An Instructional Unit: Student Text*. [New Hampshire]: New Hampshire Council for Social Studies, 1989.

667. Galarneau, Virginia. "Occupational Adjustments in the CCC Camps of Massachusetts." Master's thesis, Smith College (Mass.), 1942.

668. Galo, George. The 191st Company, The Civilian Conservation Corps: April 1939-April 1940, A Year in the Life of a Recruit. [Rutland, VT: G. Galo, 1997].

669. Gower, Calvin W. "'Camp William James': A New Deal Blunder?" *The New England Quarterly* 38 (December 1965), 475-93.

670. Hawes, Austin Foster. *The C.C.C. Makes Better Men and Forests in Connecticut*. Hartford, CT: State Forester, 1935.

671. _____. "Emergency Conservation Work in Connecticut." *Wooden Nutmeg* 9 (September 1933), 1+.

672. Judd, Richard Munson. *The New Deal in Vermont: Its Impact and Aftermath*. New York: Garland Publishing, Inc., 1979.

673. McGuire, Harvey Paul. "The Civilian Conservation Corps in Maine 1933-1942." Master's thesis, University of Maine, 1968.

674. Pasco, Mary E. "Connecticut Improves Her Trout Streams: Civilian Conservation Corps Inaugurates Pioneer Work in Developing an Important Recreation Asset." *American Forests* 40 (February 1934), 73-5.

675. Plumb, G.H. and H.W. Hicock. "Insect Control Work by the Civilian Conservation Corps in Connecticut." *Journal of Economic Entomology* 27 (April 1934), 344-5.

676. "Political Activities in the Civilian Conservation Corps: Committee Report to New England Section, Society of American Foresters." *Journal of Forestry* 31 (December 1933), 914-9; 32 (April 1934), 397-9.

677. Preiss, Jack Joseph. *Camp William James*. Norwich, VT: Argo Books, 1978.

This is the story of a group of recently graduated Harvard and Dartmouth students who wanted to make Camp William James an experimental camp in leadership training. James wanted a camp where young men of all races and creeds, from all walks of life would come together and work on projects together and learn from each other. This book tells the story of this experiment.

678. Rich, Albert T. "The CCC in New England." *Quartermaster Review* 14 (January-February 1935), 5-13.

679. Schlenker, Jon A., Norman A. Wetherington, and Austin H. Wilkins. *In the Public Interest: The Civilian Conservation Corps in Maine (1933-1942), A Pictorial History*. Augusta, ME: University of Maine Press, 1988.

This is a narrative history of CCC activities by those who served in the camps in Maine. It lists all CCC companies and gives a good description of their activities and work projects. Pictures enhance the history of each company. The appendices include a complete listing of each camp and opening and closing dates, and a map shows the location of each.

680. Sherman, Michael. "The CCC in Vermont." *Vermont History News* 45 (# 6, 1994), 74-7.

681. Stetson, Frederick W. "The Civilian Conservation Corps in Vermont." *Vermont History* 46 (Winter 1978).

682. Viers, Helen Renee. *The Civilian Conservation Corps in Vermont: A Conservation Brief.* Burlington, VT: University of Vermont, 1992.

683. Wales, H.B. "Cultural Work in the North Central States Region." *Journal of Forestry* 31 (November 1933), 766-70.

684. Webber, Edwin Waitsill. "An Evaluation of the Political History of the State of Maine: 1930-1940." University of Maine, 1937.

685. Wilson, Valerie. "The Relationship Between New Hampshire and the Civilian Conservation Corps." University of New Hampshire, 1969.

2nd Corps Area

Note: Includes New Jersey, Delaware, and New York.

686. Canton, Steve. *Dusty Roads.* Winter Park, FL: Charles B. McFadden, Co., 1993.

687. Hall, Al. "With the Civilian Conservation Corps in New Jersey." *Pennsylvania Department of Forest and Waters, Service Letter*, 29 (July 20, 1933), 2-3.

688. Hamilton, Edward A. "The Civilian Conservation Corps at Letchworth State Park, 1933-1941: A Thesis." Master's thesis, Slippery Rock University, 1984.

689. Marshall, Robert. "Facilities in Osborne's Position [re: Truck Trails]: An Open Letter to the Conservation Commissioner of New York." *The Living Wilderness* 1 (September 1935), 4-5.

690. New Jersey. Department of Institutions and Agencies. *Civilian Conservation Corps for New Jersey.* Trenton, [NJ]: New Jersey Department of Institutions and Agencies, 1939.

691. Osborne, Peter. *High Point State Park and the Civilian Conservation Corps.* Charleston , SC: Arcadia, 2002.

692. _____. *We Can Take It: The Roosevelt Tree Army at New Jersey's High Point State Park.* Bloomington, IN: 1st Books Library, 2002.

693. Potter, Barrett G. "The Civilian Conservation Corps in New York State: Its Social and Political Impact, 1933-1942." Ph.D. diss., State University of New York at Buffalo, 1973.

694. Tipper, William J. "An Educational and Personnel Study of Three Civilian Conservation Corps Camps in District II, New Jersey." Master's thesis, Temple University, 1936.

695. Verdecchio, Joseph J. "New Deal Work Relief and New York City: 1933-1938." Ph.D. diss., New York University, 1980.

696. Wilbur, C.P. "The Civilian Conservation Corps in New Jersey." *Shade Tree* 7 (March 1934), 2-4.

3rd Corps Area

Note: Includes Pennsylvania, Maryland, Virginia and the District of Columbia.

697. Avery, Myron H. "The Appalachian Trail: And the Shenandoah and Great Smokies." *The Regional Review* 4:1 (January 1940), 20-26.

698. Benson, Harvey P. "The Skyline Drive: A Brief History of a Mountain-top Motorway." *The Regional Review* 4:2 (February 1940), 3-10.

The Skyline Drive, in Virginia, was constructed by the CCC. The stone work, including the culverts, pull-offs, and entrances and exits are a good example of their work.

699. Byrne, John P. "The Civilian Conservation Corps in Virginia, 1933-1942." Master's thesis, University of Montana, 1982.

700. Conti, Philip M. *The Civilian Conservation Corps: Salvaging Boys and Other Treasures.* Harrisburg, PA: P.M. Conti, 1998.

Conti discusses his experience in CCC Co. 1317, Pine Camp S-127, in Pennsylvania.

701. Foster, Clifton Dale, comp. *Inventory of the Blue Ridge Parkway Archives.* n.p.: National Park Service, 1990.

The CCC played a major part in the construction of the Blue Ridge Parkway.

702. Eanes, Greg. *Memories of Virginia Civilian Conservation Corps Camps.* Crewe, VA: G. Eanes, 1999.
Eanes talks about prisoner of war camps at former CCC camps in Crewe, VA (Company 1370), and Green Bay, VA (Company 1390).

703. Engle, Reed L. *Everything Was Wonderful: A Pictorial History of the Civilian Conservation Corps in Shenandoah National Park.* Luray, VA: Shenandoah Natural History Association, Inc., 1999.

704. Heck, Dana B. *Phase I, Archeological Investigations at the Maintenance Complex/Park Headquarters, Fort Necessity National Battlefield, Fayette County, Pennsylvania.* Alexandria, VA: John Milner Associates, 1992.

The archaeological investigation of Fort Necessity National Battlefield included the area occupied by the CCC camp there.

705. Hendrickson, Kenneth E., Jr. "The Civilian Conservation Corps in Pennsylvania: A Case Study of a New Deal Relief Agency in Operation." *Pennsylvania Magazine of History and Biography* 100 (January 1976), 66-96.

This article discusses the organization of the CCC in Pennsylvania. It starts with a brief biographical sketch of the people involved and talks about early camps. Also discussed is the selection process for enrollees and some of the problems that arose around the state.

706. Lewis, Barbara. "A Descriptive Analysis of Civilian Conservation Corps Camp D-2 MD." Master's thesis, Salisbury State University, 1995.

707. Lotspeich, Stephen Henry. "The Design Intentions and the Planning Process of the Virginia CCC State Park Master Plans." Master's thesis, University of Virginia, 1984.

708. Neher, Leslie I. *Civilian Conservation Corps Reviewed: CCC Days of Fredericksburg Area [Virginia], 1933 Through 1937.* Gas City, IN: L.I. Neher, 1992.

709. Parker, Patricia. *The Hinterland: An Overview of the Prehistory and History of Prince William Forest Park, Virginia.* [Washington, DC]: U.S. Department of the Interior, National Park Service, National Capitol Region, [1985].

Prince William Forest Park was a Recreational Demonstration Area developed by the CCC. Recreational Demonstration Areas were parks set up for use by children from large metropolitan areas. The children could get away from the large cities and experience nature.

710. "Pennsylvania Fishing Improved by the C.C.C." *American Forests* 41 (January 1935), 41.

711. Potts, James Eugene. *Civilian Conservation Corps, Bear Valley, 1933-1936*. Chambersburg, PA: J.E. Potts, 1979.

712. Schultz, M.J. *The CCC: Its Time and Legacy*. Manitowoc, WI: Father and Son Productions, Color with B/W photographs, Sound, VHS, 25 min., 2001.

Schultz provides an overview of the CCC program, and interviews former enrollees of the Allegheny National Forest in north central Pennsylvania.

713. Simmons, Dennis Elwood. "The Creation of Shenandoah National Park and the Skyline Drive, 1924-1936." Ph.D. diss., University of Virginia, 1978.

714. Sypolt, Larry N. *Fort Necessity: CCC Camp SP-12*. Morgantown, WV: Larry N. Sypolt, 1988. (Unpublished Manuscript).

715. Vought, Robert H. "The Civilian Conservation Corps in Our State Forests." *Pennsylvania Department of Forests and Waters. Service Letter* 5 (January 18, 1934), 1-2.

4th Corps Area

Note: Includes North Carolina, South Carolina, Georgia, Florida, Alabama, Tennessee, Mississippi and Louisiana.

716. Avery, Myron H. "The Appalachian Trail: And the Shenandoah and Great Smokies." *The Regional Review* 4:1 (January 1940), 20-26.

717. Baklanoff, Joy D. *The Mounds Awaken: Mound State Monument and the Civilian Conservation Corps*. [Tuscaloosa, AL]: University of Alabama State Museum of Natural History, 1989.

Mound State Monument is a National Historic Landmark which consists of twenty mounds built by the Moundville Indians. The CCC restored the ancient Indian mounds which had been neglected and farmed over for many years. They also performed an archaeological exploration of the site, and then returned the site to its original form and contour. The CCC also constructed the museum located there.

718. Beasley, Turman Overby. "Survey of Organized Recreational Activities in Civilian Conservation Corps Camps in Tennessee." Master's thesis, Peabody University, 1935.

719. Brown, Margaret Lynn. "Smoky Mountains Story: Human Values and Environmental Transformation in a Southern Bioregion, 1900-1950 (North Carolina, Tennessee)." Ph.D. diss., University of Kentucky, 1995.

720. Buell, Jesse H. "Results of C.C.C. Timberstand Improvement on Southern Appalachian National Forests." *Journal of Forestry* 41 (February 1943), 105-12.

721. Burleson, David Albert. "Alabama's State Park System." Master's thesis, Auburn University, 1975.

722. Burns, Anna Cannaday. "The Kisatchie Story: A History of Louisiana's Only National Forest." Ph.D. diss., University of Southwestern Louisiana, 1982.

723. "C.C.C. Camps." *Mississippi Valley Lumberman* 64 (October 6, 1933), 29-30.

724. Carvalho, Joseph. "Race, Relief and Politics: The Civilian Conservation Corps in Virginia." Master's thesis, College of William and Mary, 1977.

725. Chance, William W. "Camp Roosevelt, George Washington National Forest, Page County, Va." *Evening Star* Washington, D.C. (May 1, 1933), A-3; (May 2, 1933), A-4; (May 3, 1933), A-3; (May 4, 1933), A-6; (May 5, 1933), A-6.

726. Cilbrith, Tallie. "The CCC: The Road to Recovery." *Foxfire* 16 (Winter 1982), 228-33.

This article gives insight into the CCC in South Carolina and Georgia. It looks at organization, daily work and goals of the CCC.

727. Clapp, Gordon R. *The TVA: An Approach to the Development of A Region*. Chicago: University of Chicago Press, 1955.

The CCC played a major role in the construction of the TVA dams. Several companies were located close to the dams where they cleared reservoirs and helped plan recreational areas.

728. Coleman, Beverly, R. *A History of State Parks in Tennessee.* Nashville, TN: Department of Conservation, 1968.

729. *Company 453, Florida SP-10, Civilian Conservation Corps, Sebring, Florida.* March 10, 1940. Unpublished pamphlet, Highlands Hammock State Park.

730. Foster, Clifton Dale, comp. *Inventory of the Blue Ridge Parkway Archives.* n.p.: National Park Service, 1990.

731. Gunnels, Robert D. "The Civilian Conservation Corps in Alabama." Master's thesis, Auburn University, 1990.

Gunnels presents a comprehensive study of CCC activities in Alabama. He covers the daily activities of the enrollee and lists the work projects in the state. Also, in the appendix, he provides a listing of all Alablama camps by sponsoring agency.

732. Harper, Andrew C. "The Civilian Conservation Corps and Mississippi: A New Deal Success Story." Master's thesis, University of Southern Mississippi, 1992.

733. Headrick, Leslie Mowitt and Daniel Schaffer. *A Peace Time Army: The Tennessee Valley Authority-Civilian Conservation Corps, 1933-1942.* [Norris, TN]: Tennessee Valley Authority, Division of Land and Forest Resources, Office of Natural Resources, 1983.

This report was prepared by the TVA. It is dedicated to the work performed by the CCC in TVA camps. It is heavily illustrated with photographs and maps.

734. Henry, T.S. and Irvin Foos. "Land-Conservation Camp Plan and the Tennessee River Basin." *Engineering News Record* 110 (April 6, 1933), 425-9.

This early article speculates about the proposed conservation camps that would be located in the Tennessee River Basin. It explores their possibilities in helping with projects dealing with water power, navigation and flood control, reforestation and attraction of industry.

735. Hinson, Billy G. "The Civilian Conservation Corps in Mobile County, Alabama." *Alabama Review* 45 (October 1992), 243-56.

Hinson tells of the two CCC camps in Mobile County, namely Camp Rendell and the camp near Citronelle. He tells of the work projects in each, and how they developed the natural resources and developed a recreational area in an Alabama state park.

736. Historic Property Associates, Inc. *Cultural Resource Survey: New Deal Era Resources in Nine Florida State Parks*. St. Augustine, FL: Historic Property Associates, Inc., [1989].

This monograph covers the history of nine state parks in Florida, including Florida Caverns State Park, Fort Clinch State Park, Gold Head Branch State Park, Highlands Hammock State Park, Hillsborough River State Park, Myakka River State Park, O'Leno State Park, Ravine Gardens State Park and Torreya State Park. Maps are included for each park.

737. Holmes, Michael S. "The New Deal and Georgia's Black Youth." *Journal of Southern History* 38 (August 1972), 443-60.

738. Holmes, Myrtle Falana. *Civilian Conservation Corps, Company 1420: Fort Clinch State Park*. Unpublished Manuscript., Fort Clinch State Park, Florida, 1986.

739. Hopkins, John Linn. *My Daddy was a CCC Man: I'm Proud of My Daddy: A Partial Roster of Civilian Conservation Corps Members in Mississippi in 1935 and 1937*. Memphis, TN: J.L. Hopkins, 1991.

740. Humphreys, Hubert. "In a Sense Experimental: The Civilian Conservation Corps in Louisiana." *Louisiana History* 5 (1964), 345-67.

741. _____. "In a Sense Experimental: The Civilian Conservation Corps in Louisiana." *Louisiana History* 6 (1965), 27-52.

742. Humphreys, Hubert Davis. "In A Sense Experimental: The Civilian Conservation Corps in Louisiana." Master's thesis, Louisiana State University, 1964.

743. Hurst, Fred. "Big Erosion Job: Two Thousand C.C.C. Men at Work Battling Washouts in Ten Counties in Northern Mississippi." *Forestry News Digest* (December 1933), 13.

744. Jolley, Harley E. “Southern Appalachian Forests: The last Fifty Years.” *The Great Forest: An Appalachian Story.* Boone, [WV]: Appalachian Consortium Press, 1985, 23-8.

745. Jolley, Harley E., Steve Kemp and Kent Cave. *The CCC in the Smokies.* Gatlinburg, TN: Great Smoky Mountains Natural History Association, 2001.

746. Jones, James Boyd, Jr. "Civilian Conservation Corps: Its Legacy to Tennessee." *The Courier* 22 (June 1984), 4-7.

747. Jones, James B. and Claudette Stager. *State Parks in Tennessee Built by the CCC and WPA Between 1934-1942.* Unpublished National Register of Historic Places Multiple Property Documentation, Nashville, TN: Tennessee Historical Commission, 1986.

748. Kennerly, Edna. "The Civilian Conservation Corps as a Social Resource in South Carolina." Master's thesis, University of South Carolina, 1940.

749. Lawliss, Lucy Ann. "The Civilian Conservation Corps and the State Park: An Approach to the Management of the Designed Historic Landscape Resources at Franklin D. Roosevelt State Park, Pine Mountain, Georgia." Master's thesis, University of Georgia, 1992.

750. Lenhart, D.Y. *The South Carolina Civilian Conservation Corps Forester.* Columbia, SC: State Commission of Forestry, 1936.

This is an informative little forestry pamphlet. It is fully illustrated and meant as a teaching and training aid for those enrollees working on reforestation projects. It provides situations and possible solutions. It also lists supervisory and foreman personnel lists for each forestry camp, in the front of the pamphlet.

751. Lentz, Gus. "Making Water Walk in the Tennessee Valley." *American Forests* 40 (May 1934), 200-2+.

752. Manchester, E.R. "Our Fugitive Farm Lands: Let’s Move up to the Front Where CCC Workers in Mississippi are Using Explosives to Check Soil Erosion." *DuPont Magazine* 28 (February 1934), 1-3.

753. Martin, Roscoe C., ed. *TVA: The First Twenty Years, A Staff Report.* Knoxville, TN: The University of Alabama Press and The University of Tennessee Press, 1956.

754. Mastran, Shelley Smith. *Mountaineers and Rangers: A History of the Federal Forest Management in the Southern Appalachians, 1900-81.* Washington, DC: USDA Forest Service, 1983.

Mastran devotes two chapters of this book to the New Deal and CCC in Kentucky, Tennessee, North Carolina, South Carolina and Georgia. A chart tells how many resident and non-resident enrollees were employed in these Appalachian states in 1934, 1937 and 1940.

755. McNew, Robert Lyon. "A Study of the Civilian Conservation Corps with Special Emphasis on the South." Master's thesis, Austin Peay State University, 1971.

756. National Association for the Advancement of Colored People. *NAACP Legal File: Cases Supported, CCC Boys, 1938-1939.* Frederick, MD: University Publications of America, 1988.

This entry covers correspondence, memoranda and press releases concerning the cases of three Black CCC enrollees charges with murder in Georgia.

757. Nelson, David J. *Relief and Recreation: The Civilian Conservation Corps and the Florida Park Service, 1933-1942.* Master's thesis, Florida State University, 2002.

758. North Carolina Board of Charities and Public Welfare. *Biennial Report of the State Board of Charities and Public Welfare, July 1, 1938 to June 30, 1940.* {Chapel Hill, NC}: Academic Affairs Library, University of North Carolina at Chapel Hill, 2003.

One section of this report is titled “Selection and Certification of Applicants for Civilian Conservation Corps.”

759. The Official Annual, Civilian Conservation Corps, "C" District. CCC Fourth Corps Area. Baton Rouge, LA: Direct Advertising Company, July 1937.

760. Sadka, George Norman. "Mississippi Selection Service for Youths in the Civilian Conservation Corps." Master's thesis, Tulane University of Louisiana, 1948.

761. Shofner, Jerrell H. "Roosevelt's 'Tree Army:' The Civilian Conservation Corps in Florida." *Florida Historical Quarterly* 65 (April 1987), 433-56.

762. "South Given Additional C.C.C. Camps." *American Forests* 39 (December 1933), 563.

763. Sternsher, Bernard, ed. *Hope Restored: How the New Deal Worked in Town and Country.* Chicago: Ivan R. Dee, 1999.

Billy G. Hinson writes an article "Civilian Conservation Corps in Mobile County, Alabama, in this book. The book deals with social and political issues of the time.

764. "Summary by States of C.C.C. Work Shows Great Advance Made in Forest Program: Much Fire Fighting." *Forestry News Digest* (November 1934), 8-12.

765. Sumner, Jim L. "The Civilian Conservation Corps and the Development of North Carolina's State Park System." *Carolina Comments* 40 (July 1992), 98-106.

766. Sweets, John F. "The Civilian Conservation Corps in Florida." *Apalachee* VI (1963-1967).

767. Tate, Roger D., Jr. "George D. Power and New Deal Work Relief in Mississippi, 1933-1934." *Journal of Mississippi History* 46:1 (1984), 1-16.

768. Waller, Robert A. "Happy Days and the Civilian Conservation Corps in South Carolina, 1933-1942." *The Historian; A Journal of History* 64:1 (2001), 39-63.

769. Wimbush, Solomon Mitchell. "Louisiana, the Civilian Conservation Corps Black Camps in District 'E': 1933-1942." Master's thesis, Louisiana State University and Agricultural and Mechanics College, 1972.

5th Corps Area

Note: Includes Ohio, West Virginia, Indiana and Kentucky.

770. Anderson, Robert E. *Written on the Land.* Fairmont, WV: Robert E. Anderson, 2002.

Anderson provides a pictorial history of the CCC in West Virginia.

771. Beanblossom, Robert. “Civilian Conservation Corps: The Early Foundation of West Virginia’s State Parks.” *Wonderful West Virginia* 64 (April 2000), 20-24.

772. Bickley, Ancella. “Camp War: Remembering CCC Company 3538-C.” *Goldenseal* 27 (winter 2001), 22-9.

773. Carvell, Kenneth L. *Camp Rhododendron: Civilian Conservation Corps Camp at Cooper’s Rock State Forest (1936-42).* Morgantown, WV: Monongalia Historical Society, 1996.

This short booklet was prepared for the 60th anniversary of the opening of Camp Rhododendron. It provides a history and photographs of Company 3527.

774. Casto, James E. “Civilian Conservation Corps.” *Wonderful West Virginia* 61 (February 1998), 4-7.

This article discusses work in several state parks in West Virginia.

775. Chancellor, George. "The CCC at Fort Knox." *Cavalry Journal* 42 (July-August 1933), 36+.

776. Comstock, Jim. "How Come Richwood?" *Wonderful West Virginia* 48:10 (December 1983), 14-17.

Tells how Richwood, WV, was chosen for the 50th anniversary reunion of former CCC enrollees from West Virginia.

777. Daugherty, Raymond. *The CCC and Me.* Mathias, WV: Raymond Daugherty, 2003.

Daugherty talks about social conditions in West Virginia before joining the CCC. He continues with his work in CCC Co. 1524 on Lost River State Park, WV.

778. Dunnington, Garrett. “Daily Life in the Tree Army: The CCC Diary.” *Muskingum Journal* 10:2 (Winter 2003), 7-22.

This article deals with the CCC Camp at Camp Wyoming in Pineville, WV. Included are pages from Dunnington’s journal from 1935.

779. Farley, Yvonne S. "'A Good Part of Life' Remembering the Civilian Corps [sic]" and "The Civilian Conservation Corps and Babcock State Park." *Goldenseal* (January-March 1981).

780. Harr, Milton. *The C.C.C. Camps in West Virginia: A Record of the Civilian Conservation Corps in the Mountain State.* Charleston, WV: Milton Harr, 1992.

This booklet provides a chronological list of all CCC companies and camps in West Virginia along with some of their accomplishments.

781. Headington, L.H. "A Study of the Duties of Company Commanders, Educational Advisers, and Project Superintendents in the Civilian Conservation Corps of West Virginia." Master's thesis, Ohio State University, 1938.

782. Hersel, George. "Legacy of the CCC." *Wonderful West Virginia* 47 (December 1983), 3-11.

783. Leffler, Susan. "Nothing But Just Fighting: The 1936 CCC Race Riot." *Goldenseal* 27 (Winter 2001), 30-3

This article is based on an oral interview with Bill Dougherty, a black enrollee at Camp Loring near Minnehaha Springs, WV. Dougherty tells of a "race riot" which broke out over Black enrollees refusing to work because of severe weather conditions, and repercussions by white officers.

784. Mastran, Shelley Smith. *Mountaineers and Rangers: A History of the Federal Forest Management in the Southern Appalachians, 1900-81.* Washington, DC: USDA Forest Service, 1983.

785. McBride, Dennis. *Hard Work and Far From Home: The Civilian Conservation Corps at Lake Mead, Nevada.* Boulder City, NV: Boulder Images, 1995.

Many enrollees on this project were from the 5th Corps Area. They were transferred out west to work at Lake Mead. This personal recollection includes quotes and stories from many enrollees who were actually there.

786. Sessions, Elden B. "Educational Work of the CCC Camps in Ohio." Ph.D. diss., Ohio State University, 1937.

787. Smith, Kathy Mays. "Army Navy Journal." and "Gold Medal CCC Camp." *Armed Forces Journal International* (August 2001), 25-6, 28-31.

Smith looks at the early years of the CCC and the award given to the best company in each of the nine Army Corps Areas. The best non company commanders were sent to Washington to receive medals from General Douglas MacArthur.

788. _____. *Gold Medal CCC Company 1538: A Documentary.* Paducah, KY: Turner Publishing Co., 2001.

Smith tells of her father's long military career, from commanding a CCC Company (Co. 1538) in Pineville, WV to becoming the Marshall of the Court at the Nurnberg War Tribunal. Capt. Mays' Co. 1538 was one of the first nine companies to be recognized as the most outstanding company in its respective Corps Area. He received a gold medal from Gen. Douglas MacArthur for his efforts.

789. Taylor, John Craft. "Depression and New Deal in Pendleton: A History of a West Virginia County from the Great Crash to Pearl Harbor, 1929-1941." Ph.D. diss., Pennsylvania State University, 1980.

In part, he discusses how this West Virginia county benefitted from having CCC camps located in both the Monongahela and George Washington National Forests within its boundaries. He mentions camp locations and work projects.

790. "Voluntary Political Contributions to Indiana." *Journal of Forestry* 32 (November 1934), 874-6.

791. Walker, Helen Mabel. *The C.C.C. Through the Eyes of 272 Boys: A Group Study of the Reactions of 272 Cleveland Boys to their Experience in the Civilian Conservation Corps.* Cleveland, OH: Western Reserve University Press, 1938.

792. Ward, Ernest H. "A Study of the Leisure-Time Activities of the Civilian Conservation Corps in Ohio." Master's thesis, Ohio State University, 1937.

6th Corps Area

Note: Includes Illinois, Michigan and Wisconsin.

793. Ahlgren, Carol. "The Civilian Conservation Corps and Wisconsin State Park Development." *Wisconsin Magazine of History* 71 (Spring 1988), 184-202.

Ahlgren writes a good history of CCC involvement in Wisconsin state parks. She tells the history of Wisconsin parks and the influential people involved in their development before and during the CCC years. The article includes many photographs of CCC construction projects.

794. _____. "A Human and Landscape Architectural Legacy: The Influence of the Civilian Conservation Corps on Wisconsin State Park Development." Master's thesis, University of Wisconsin-Madison, 1987.

795. Elliott, Kenneth M. *History of the Nicolet National Forest 1928-1976.* Forest History Assn. Of Wisconsin and Forest Service, 1977.

Chapter Four of this publications discusses the role of the CCC in Nicolet National Forest. It lists all of the CCC camps in the forest, their locations and and some of the work projects.

796. Hubbard, Nancy J., et al. *1991 State Facilities Survey: Reconnaissance Survey Report.* Milwaukee, WI: University of Wisconsin-Milwaukee, School of Architecture and Urban Planning, [1992].

797. Mance, Frank. "An Historical Study of the Civilian Conservation Corps in Illinois." Master's thesis, Western Illinois University, 1967.

798. Raff, Willis H. *CCC Days in Cook County, 1933-1942.* Grand Marais, MN: Cook County Historical Society, 1983.

799. Reich, Edwin W. *Civilian Conservation Corps, Camp Ellsworth, Wisconsin, Company #3669-SCS-3, 1935-1942.* Ellsworth, WI: The Company, 2001.

800. Rosentreter, Roger L. "Michigan's 83 Counties: Oscoda County." *Michigan History* 73:4 (1989), 8-10.

801. _____. "Roosevelt's Tree Army: The Civilian Conservation Corps in Michigan." *Michigan History* 70 (May/June 1986), 14-23.

802. _____. *Roosevelt's Tree Army: Michigan's Civilian Conservation Corps.* [Lansing, MI]: Michigan Department of State, 1986.

803. Semo, John W. *The Story of Camp Pattison.* [Madison, WI?]: Wisconsin Department of Natural Resources, 2003.

This pamphlet well of the work done by CCC Company 3663 in Pattison State Park.

804. Sorenson, Gordon. "The Civilian Conservation Corps in Wisconsin." *Timber Producer* (March 1977), 26-9.

805. Symon, Charles A. *We Can Do It! A History of the Civilian Conservation Corps in Michigan--1933-1942.* Escanaba, MI: Richards Printing, 1983. Symon brings all aspects of the history of the Michigan CCC together in his book. It covers the various types of camps including veteran and Native American, and work projects which are enhanced with maps. The final pages include a short history of the individual camps, a listing of all of the camps

7th Corps Area

Note: Includes Kansas, Arkansas, Iowa, Nebraska, Minnesota, North Dakota, South Dakota and Missouri.

806. Alleger, C.N. *Civilian Conservation Corps, Minnesota District: That the Work of the Young May Be Recorded.* n.p.: Johnston and Bordewyck, Inc., 1934.

807. Anderson, Rolf T. *Minnesota State Park CCC/WPA/Rustic Style Historic Resources.* Minneapolis, MN: R.T. Anderson, 1988.

808. Bauer, Merle. "A Survey of Civilian Conservation Corps Contributions to Recreation in Nebraska." University of Wyoming, 1956.

809. Benson, David R. *Stories in Log and Stone: The Legacy of the New Deal in Minnesota State Parks.* St. Paul, MN: Minnesota Department of natural Resources, Division of Parks and Recreation, 2002.

A good deal of this book is dedicated to the the CCC in Minnesota. It talks about how the New Deal got organized, the work projects of the CCC, and the development of parks and park structures.

810. "Company 5757, CCC, by the Company Commander." *Saturday Evening Post* 210 (September 11, 1937), 8-9+.

811. Derschied, Lyle A. *The Civilian Conservation Corps in South Dakota (1933-1942): A Historical Report Sponsored by Chapter 61 of NACCCA.* Brookings, SD: SDSU Foundation Press, 1991.

812. Gaffney, Rebecca and Mary Ann Quint. *History of Backbone: Iowa's First State Park.* Manchester, IA: Iowa Department of Natural Resources, State Parks Northeast District Office, 2002.

Gives a history of the two CCC Companies that helped build Backbone State Park, Company 781, Lamont, IA, and Company 1756, Dundee, IA.

813. Giesler, Cecil O. "How I Plan to Use Experience Acquired in an S.C.S.-C.C.C. Camp [in Missouri]." *Soil Conservation* 3 (March 1938), 242+.

814. Haus, Lawrence Harry. "A Study of the Civilian Conservation Corps with a Mental Survey of Camp Lake Wilhite, Reading, Kansas." Master's thesis, Kansas State Teachers College of Emporia, 1935.

815. Hendrickson, Kenneth E., Jr.. "The Civilian Conservation Corps in South Dakota." *South Dakota History* 11 (Winter 1980), 1-20.

816. _____. "Relief for Youth: The Civilian Conservation Corps and the National Youth Administration in North Dakota." *North Dakota History* 48 (Fall 1981), 17-27.

817. Koncker, Janelle. *A Living History: An Interactive Learning Museum About the CCC in Northeast Minnesota.* MN: Typescript & CD-ROM, 2003.

818. Long, Fred H. "Two Decades of State Forestry in Arkansas." *Arkansas Historical Quarterly* 24:3 (1965), 208-19.

819. McCarty, Joey. "Civilian Conservation Corps in Arkansas." Master's thesis, University of Arkansas, Fayetteville, 1977.

820. McClurkan, Burney B. *A Brief History of Camp Sage: CCC Camp P-76, Stella, Arkansas.* Little Rock, AR: Arkansas State Highway and Transportation Department, 1994.

821. Murphy, James Lee. "A History of the Southeastern Ozark Region of Missouri." Ph.D. diss., Saint Louis University, 1982.

822. Nelson, Edward P. and Barbara Sommer. *It was a Good Deal: The Civilian Conservation Corps in Northeastern Minnesota; A Guide to the Arrowhead Civilian Conservation Corps Document Project.* Duluth, MN: St. Louis Historical Society, 1987.

823. O'Dell, Charles A. *The Civilian Conservation Corps in Arkansas, 1933-1934.* Columbia, MO: C.A. O'Dell, 1995.

824. Shunk, Harold W. (Mato Ska) "Reminiscing About the Dakota." *Kansas Quarterly* 3:4 (1971), 116-23.

825. Smith, Sandra Taylor. *The Civilian Conservation Corps in Arkansas, 1933-1942: A Historic Context.* Little Rock, AK: Arkansas Historic Preservation Program, 1991.

826. Thoma, Ben. *The Civilian Conservation Corps and Itasca State Park.* Lake Itasca, MN: C.C.C. History Project, Itasca State Park Headquarters, 1984.

827. Wilson, Heather M. *A Brief History of the National Park Service Civilian Conservation Corps Camps in Nebraska.* n.p., 1985.

8th Corps Area

Note: Includes Texas, Oklahoma, Colorado, New Mexico, Arizona and Wyoming (less Yellowstone National Park).

828. Alleger, C.N. and L.A. Gleyre, comps. *History of the Civilian Conservation Corps in Colorado.* Denver, CO: Western Newspaper Union, 1936.

829. Anderson, Rolf T., Susan Granger, Scott Kelly, and Kay Grossman. *Federal Relief Construction in Minnesota, 1933-1941.* MN: Gemini Research, 2001.

Section E. II, covers the role of the Civilian Conservation Corps in Minnesota. This is part of a National Register of Historic Places Multiple Property Documentation form.

830. Baker, Robert D., Robert S. Maxwell, Victor H. Treat, and Henry C. Dethloff. *Timeless Heritage: A History of the Forest Service in the Southwest.* USDA Forest Service, 1988.

831. Boeger, Palmer H. *Oklahoma Oasis: From Platt National Park to Chickasaw National Recreation Area.* Muskogee, OK: Western Heritage Books, 1987.

832. Booth, Peter MacMillan. "Cactizonians: The Civilian Conservation Corps in Pima County, 1933-1942." *Journal of Arizona History* 32 (Autumn 1991), 291-332.

833. _____. "The Civilian Conservation Corps in Arizona, 1933-1942." Master's thesis, University of Arizona, 1991.

Booth's thesis provides a good history of the CCC, regular and Indian Division, in Arizona. He talks about the people responsible for the work projects, educational activities and training. Included is material on Hispanics in Arizona and how their association with the CCC promoted their insight into American society.

834. Briscoe, James M. *The Civilian Conservation Corps at Possum Kingdom State Park, Palo Pinto County, Texas.* [Norman, OK?}: Briscoe Consulting Services, 2000.

835. Carney, Elizabeth. *The Civilian Conservation Corps on the San Juan National Forest.* Durango, CO: San Juan National Forest Service, 1994.

836. Cassady, John Thomas. *Revegetating Semidesert Range Lands in the Southwest.* Washington, DC: Government Printing Office, 1940.

837. Clapp, Gordon R. *The TVA: An Approach to the Development of A Region.* Chicago: University of Chicago Press, 1955.

838. Clark, J.D. "CCC in the Southwest." *Industrial and Engineering Chemistry News Edition* 12 (July 10, 1934), 249-50.

839. Cockrum, E. Lendell. *Cavemen of Colossal Cave.* Typescript by E. Lendell Cockrum, 2000.

History of CCC Co. 2851 at Vail, Arizona.

840. Copenhaver, Larry. *CCC Camps in Arizona.* n.p.: n.p., 1966.

841. Daubner, Nathan. "The History of the WPA and the CCC in Winona County [MN]." Historical Research Paper, St. Mary's College, 1993.

842. Gart, Jason H. *A Narrative History of the Civilian Conservation Corps at Chiricahua National Monument, Cochise County, Arizona.* Mesa, AZ: History International, 2001.

843. Gimmerson, Johanna K. *German POW Camp in Deaver, Wyoming, 1944-1945.* Powell, WY: J.K. Gimmerson, 2001.

844. Gleyre, L.A. and C.N. Alleger. *History of the Civilian Conservation Corps in Colorado.* Denver, CO: Press of the Western Newspaper Union, 1936.

845. _____. *History of the Civilian Conservation Corps in Colorado, Littleton District-Grand Junction District: That the Work of Young America May Be Recorded.* Denver, CO: n.p., 1936.

846. Hanneman, Carolyn G. "Baffles, Bridges, and Bermuda: Oklahoma Indians and the Civilian Conservation Corps–Indian Division." *Chroni cles of Oklahoma* 77:4 (Winter, 1999), 428.

847. Hanson, James Austin. "The Civilian Conservation Corps in the Northern Rocky Mountains." Ph.D. diss., University of Wyoming, 1973.

Hanson gives a comprehensive overview of the CCC program and its history. He includes many of the CCC camp designation numbers and locations for Idaho, Montana and Wyoming.

848. Holland, Reid A. "The Civilian Conservation Corps in the City: Tulsa and Oklahoma City in the 1930s." *Chronicles of Oklahoma* 53 (Fall 1975), 367-75.

849. _____. "Life in Oklahoma's Civilian Conservation Corps." *Chronicles of Oklahoma* 48 (Summer 1970), 224-34.

An abbreviated article based on the following Master's thesis.

850. Holland, Reid Allen. "The Civilian Conservation Corps in Oklahoma." Master's thesis, Oklahoma State University, 1969.

851. Kimmett, Leo. "Life in a Yellowstone CCC Camp." *Annals of Wyoming* 56 (Spring 1984), 12-21.

852. Koncker, Janelle. *A Living History: An Interactive Learning Museum About the CCC in Northeast Minnesota."*

853. Lindley, J.G. "Protecting Colorado's Range Lands." *Soil Conservation* 1 (April 1936), 12-3.

854. _____. "Soil Conservation and the C.C.C." *Soil Conservation* 2 (May 1937), 284-5.

855. Malach, Roman. *Home on the Range: Civilian Conservation Corps in the Kingman Area.* Kingman, AZ: R. Malach, 1984.

856. Melzer, Richard. *Coming of Age in the Great Depression: The Civilian Conservation corps Experience in New Mexico, 1933-1942.* Las Cruz, NM: Yucca Tree Press, 2000.

857. Mergen, Bernard. "Denver and the War on Unemployment." *Colorado Magazine* 47:4 (1970), 326-337.

858. Montoya, Maria E. "The Roots of Economic and Ethnic Divisions in Northern New Mexico: The Case of the Civilian Conservation Corps," *Western Historical Quarterly* (Spring 1995), 14-34.

This article compares the situation of Native Americans and Hispanics to that of Anglos in Northern New Mexico. While Native Americans served in separate camps, the Hispanics were integrated with Anglos in camps, and good records were not kept on them. It discusses the lack of concern for and understanding of social and economic culture of the minorities by administrators.

859. Parham, Robert Bruce. "The Civilian Conservation Corps in Colorado, 1933-1942." Master's thesis, University of Colorado, Boulder, 1981.

860. Parr, Lewis A. "A Personnel Study of the School Achievement, Intelligence Socio-Economic Status, and Personality Adjustment of Civilian Conservation Corps Enrollees of Northeast Texas." Master's thesis, University of Texas, 1937.

861. Pearson, Gustaf Adolph. *Reforestation in the Southwest by CCC Camps*. Washington, DC: Government Printing Office, 1940.

862. Purvis, Louis Lester. *The Ace in the Hole: A Brief History of Company 818 of the Civilian Conservation Corps*. Columbus, GA: Brentwood Christian Press, 1989.

863. Richardson, Elmo R. "The Civilian Conservation Corps and the Origins of the New Mexico State Park System." *Natural Resources Journal* 6 (April 1966), 248-67.

864. Sallee, Katherine Northcut. "The Cultural Landscape Inventory: A Research Process for the Platt District of the National Park Service (Oklahoma)." Master's thesis, University of Texas at Arlington, 1997.

865. Schrems, Suzanne H. "A Lasting New Deal Legacy: The Civilian Conservation Corps, The National Park Service and the Development of the Oklahoma Park System." *Chronicles of Oklahoma* 72 (Winter 1994-95), 369-95.

866. Scrattish, Nicholas. *Historic Resource Study, Bryce Canyon National Park*. Denver, CO: National Park Service, 1985.

867. Seton, Max C. *Canadian River State Park, Woodward, Oklahoma.* Washington, DC: National Park Service, 1987.

868. Seymour, Gregory R. "The Civilian Conservation Corps in Arizona: A Context for Erosion-Control Features." *The Kiva* 63 (1998).

In this article, Seymour discusses the methods of soil erosion control used by the various agencies supervising the CCC enrollees in Arizona. He discusses, among others, rock spreaders, gabions, cement worm spreaders rock and brush spreaders and diversion dams. Also included is a map showing the locations of the CCC camps in Arizona and the number of camps in each enrollment period.

869. _____. *The Sanchez Copper Project: The Civilian Conservation Corps at the Sanchez Camp, SCS-11-A: A Camp History and Architectural Evaluation, Graham County, Arizona, Vol. 2.* SWCA Archaeological Report No-95-13, Tucson, AZ.

870. Sherard, Gerald Emerson. *Colorado Civilian Conservation Corps Enrollment.* Lakewood, CO: G.E. Sherard. 1997.

871. Smith, Monica L. "The Archaeology of a 'Destroyed' Site: Surface Survey and Historical Documents at the Civilian Conservation Corps Camp, Bandelier National Monument, New Mexico." *Historical Archaeology* 35:2 (2001), 31-40.

872. Steely, James Wright. *The Civilian Conservation Corps in Texas State Parks.* Austin, TX: Texas Parks and Wildlife Department, 1986.

873. _____. "Rustic Style in Depression Texas: Federal Architecture in the State Parks, 1933-1941." Master's thesis, University of Texas at Austin, 1985.

874. Thomas, Karen. *Camp Arboretum: The Dynamics of the Civilian Conservation Corps in McDonald Forest and Surroundings [sic] Areas.* Albuquerque, NM: Heritage Associates, 1981.

875. Thurman, Claude Harrison. "The Curricular and Instructional Program of the Junior White Civilian Conservation Corps Camps of the North and South Texas Districts." Ph.D. diss., University of Texas, 1940.

876. Touchet, Robert W. "New Deal Work Programs in Jefferson County, Texas: The Civilian Conservation Corps at Tyrell Park." Lamar University, 1972.

877. Utley, Dan K., James Wright Steely and T. Lindsay Baker. *Guided with a Steady Hand: The Cultural Landscape of a Rural Texas Park.* Waco, TX: Baylor University Press, 1998.

878. Van Norden, Maureen. "Civilian Conservation Corps Guidelines for Restoration and Preservation of Colorado." Master's thesis, University of Colorado at Denver, 1989.

879. Weber, R.C.E. "Emergency Conservation Work, Yuma Project, Arizona-California." *Reclamation Era* 26 (May 1936), 124-5.

880. Wellborn, Mark Alan. "Texas and the CCC: A Case Study in the Successful Administration of a Confederated State and Federal Program." Master's thesis, University of North Texas, 1989.

881. Wilson, R.E. "Civilian Conservation Corps Work in the South-Western United States." *Forestry Chronicle* 11 (December 1935), 317-22.

882. Wisenberger, Carol Whiteside. The National Youth Administration in Texas, 1935-1943: A Case Study." Ph.D. diss., Texas A & M University, 1988.

9th Corps Area

Note: Includes Washington, Oregon, Idaho, Montana, Utah, Nevada, California and Yellowstone National Park (in Wyoming).

883. Arrington, Leonard. "The New Deal in the West: A Preliminary Statistical Inquiry." *Pacific Historical Review* 38 (August 1969), 311-6.

884. Austin, Judith. "The CCC in Idaho." *Idaho Yesterdays* 27 (Fall 1983), 13-8.

885. Baldridge, Kenneth W. "Nine Years of Achievement: The Civilian Conservation Corps in Utah." Ph.D. diss., Brigham Young University, 1971.

886. Bergeson, Thora. "A Survey of the Social Background of the Civilian Conservation Corps Enrollees of the Fort Douglas District." Master's thesis, University of Utah, 1937.

887. Barber, Lawrence. "Tying Together Oregon's Mountains: A Network of New Forest Roads is Being Thrown Over Oregon's Mountains by the Civilian Conservation Corps." *Oregon Motorist* (July 1934), 7-8+.

888. Barden, Harold E. "A Social Adjustment of California Civilian Conservation Corps Enrollees as Determined by a Critical Analysis of Their School, Camp and Post-Camp Records." Ph.D. diss., University of Southern California, 1939.

889. Burke, Robert E. *Olson's New Deal for California.* Berkeley, CA: University of California Press, 1953.

890. Carew, Harold D. "A Fair Deal Wins–A Foul Dole Loses." *California History* 63 (Fall 1983), 172-4.

Carew gives a short overview of the CCC in California. He tells of the work program started by the state of California for the large unemployed transient population before December 1931. Several CCC camps in the Ninth Corps Area are listed.

891. Carroll, Robert Wesley. "The Civilian Conservation Corps in Washington State, 1933-1942." Master's thesis, Washington State University, 1973.

892. "C.C.C. Camp Lewis and Clark" *Timberman* 35 (December 19, 1933), 104.

893. Clark, J.D. "CCC in the Southwest." *Industrial and Engineering Chemistry News Edition* 12 (July 10, 1934), 249-50.

894. Cox, Thomas R. *The Park Builders: A History of State Parks in the Pacific Northwest.* Seattle, WA: University of Washington Press, 1988.

Cox provides a history of the state park systems in Idaho, Oregon and Washington. He provides information on individuals responsible for the state park movements in each state, provides maps of the locations of state and national parks in these three states, and talks about CCC construction in them..

895. Crosby, Michael. *"I'd Do It Again in a Minute!" The Civilian Conservation Corps on the Salmon National Forest.* Salmon, ID: M. Crosby, 1997.

896. Deering, R.L. "Camps for the Unemployed in the Forests of California." *Journal of Forestry* 30 (May 1932), 554-7.

897. Fairley, Lincoln. *Mount Tamalpais: A History.* San Francisco, CA: Scottwall Associates, 1987.

898. Haymond, Jay Melvin. "History of the Miami Forest, Utah: A Case of Conservation in the West." Thesis, University of Utah, 1972.

899. Hinton, Wayne. "The New Deal Years In Utah: A Political History of Utah (1932-1940)." Master's thesis, Utah State University, 1963.

900. Housley, Harold. "Notes and Documents: Elwood Decker and the CCC at Fort Churchill." *Nevada Historical Society Quarterly* 38 (Summer 1995), 105-21.

Decker was an artist who documented the restoration and reconstruction of Fort Churchill in his art and writings. This article includes many examples of the art work he produced on Fort Churchill.

901. Hussey, W.L. "Brief History of Company 957, CCC." (n.p., pamphlet).

902. Hutchinson, W.I. *Public Opinion of the Civilian Conservation Corps in California.* San Francisco: U.S. Forest Service, California Region, 1934. (Pamphlet)

903. James, Charles D. III. *California Labor Camps.* Chico, CA: n.p., 1988.

904. Jewett, Stephanie. "The Civilian Conservation Corps in Iron County and Surrounding Areas." Senior thesis, Southern Utah University, 1994.

905. Karsmizki, Kenneth W. "The Lewis and Clark Caverns: Politics and the Establishment of Montana's First State Park." *Montana* 31:4 (1981), 32-45.

906. Ketchum, Smith A. "Civilian Conservation Corps Activities on the Kennett Division, Central Valley Project." *Reclamation Era* 29 (April 1939), 85-6.

907. Kimmett, Leo. "Life in a Yellowstone CCC Camp." *Annals of Wyoming* 56 (Spring 1984), 12-21.

908. Lowitt, Richard. *The New Deal and the West.* Bloomington, IN: Indiana University Press, 1984.

909. Maxey, William, Jr. "Civilian Conservation Corps Activities on the Milk River Project, Montana." *Reclamation Era* 29 (October 1939), 257-9

910. McBride, Dennis. *Hard Work and Far From Home: The Civilian Conservation Corps at Lake Mead, Nevada.* Boulder City, NV: Boulder Images, 1995.

911. Mueller, Edward. "What the Civilian Conservation Corps Means to 115 Los Angeles Boys who were Honorably Discharged in March, 1940." Master's thesis, University of Southern California, 1941.

912. Neef, Diana L. "Civilian Conservation Corps: Elko County, 1933-1942." *Northeastern Nevada Historical Society Quarterly* (1984), 16-24.

913. Nelsen, Clair E. "Remembering the CCC: Buck Meadows, California, 1933-1934." *Journal of Forest History* 26 (October 1982), 184-91.

914. Nelson, Neal McMaster. "An Historical Inquiry in the Civilian Conservation Corps Movement, with Special Reference to the Ninth Corps Area." Master's thesis, University of Idaho, 1938.

915. Ober, Michael J. "The CCC Experience in Glacier National Park." *Montana: The Magazine of Western History* 26 (July 1976), 30-9.

916. Olsen, Beth R. "Utah's CCCs: The Conservators' Medium for Young Men, Nature, Economy, and Freedom." *Utah Historical Quarterly* 62:3 (1994), 261-74.

917. Parkinson, Dana. "The CCC Helps Farmers in Utah." *The Utah Farmer* (March 25, 1935), 10.

918. Patterson, James R. "The New Deal in the West." *Pacific Historical Review* 38 (August 1969), 317-27.

919. Pelley, Albert Louis. "An Investigation of Educational Opportunities in the Civilian Conservation Corps for the Ninth Corps Area." Master's thesis, University of Wyoming, 1941.

920. Pinkerton, John B. "The Youth Development and Conservation Corps in the State of Washington." *Journal of Forestry* 62 (June 1964), 373-5.

921. Reichert, Bruce and Pat Metzler. *Outdoor Idaho. CCC in Idaho.* Boise, ID: Idaho Public Television, Color, Sound, VHS, 30 min., 1999.

922. Sacco, Rose. "Civilian Conservation Corps in Shasta County, 1933-1942." Term Paper, Shasta College (Redding, CA), 1969.

923. Santino, Mathew. *Civilian Conservation Corps, 1933-1937; [Materials Pertaining to the History of the Corps in Humboldt County, California]*.

924. Savage, Christine E. *New Deal Adobe: The Civilian Conservation Corps and the Reconstruction of Mission La Purisima, 1933-1942.* Santa Barbara, CA: Fithian Press, 1991.

Savage tells of the reconstruction of Mission La Purisima by the CCC. She takes much of her information from the mission archives and presents a good personal description of supervisors and enrollees and their dedication to an authentic reconstruction as possible.

925. Shinn, Paul L. "Eugene in the Depression, 1929-1935." *Oregon Historical Quarterly* 86:4 (1985), 341-69.

926. Smith, M.L. "Evenings at Cuyamaca." *Sierra Education News* 36 (February 1940), 24+.

927. Supernowicz, Dana E. *Contextual History of Forest Service Administration Buildings in the Pacific Southwest.* USDA Forest Service, Region 5, California, 1989.

Among other things, this is a history of the design and construction of Forest Service buildings by architects E.J. Maher and N. Blanchard, and the influence they had on future construction. They used native and ready-cut materials in the Craftsman-Bungalow designs of the 1920's and the California "ranch" style of the 1930's.

928. Thomas, Margaret E. *The Civilian Conservation Corps Program in the State of Washington, April, 1933-October, 1936* Olympia, WA: State of Washington Department of Social Security (Monograph no. 23), 1937.

929. Throop, Elizabeth Gail. "Utterly Visionary and Chimerical: A Federal Response to the Depression: An Examination of Civilian Conservation Corps Construction on National Forest System Lands in the Pacific Northwest." Master's thesis, Portland State University, 1979.

Throop's thesis presents an inventory of extant Depression era sites and structures in national forests in the Pacific Northwest (Oregon and Washington). A good overview of CCC operations is followed by a listing of forests being inventoried. This thesis contains maps of both Washington and Oregon pinpointing where administrative and recreational structures and sites were located.

930. Trentman, Mark Daniel. "One Dollar A Day Men: The Civilian Conservation Corps on the North Coast of California, 1933-1942." Senior thesis, Humboldt State University, 1981.

931. *Walcott Park.* Loveland, CO: Fraserdesign, 1998, (Historic American Buildings Survey, HABS No. ID-103; Minidoka County, Idaho).

This is a HABS documentation of Walcott Park which served as a construction and operations camp for Minidoka Dam and Powerplant. CCC work projects included landscaping, tree planting, placement of embankment stone, and construction of retaining walls and parking areas.

932. Weber, R.C.E. "Emergency Conservation Work, Yuma Project, Arizona-California." *Reclamation Era* 26 (May 1936), 124-5.

933. White, William G. *The Civilian Conservation Corps in Clark County, Nevada: An Historic Overview and Context.* Las Vegas, NV: Harry Reid Center for Environmental Studies, University of Nevada, Las Vegas, 2003.

AFRICAN AMERICANS, NATIVE AMERICANS & MINORITIES

African Americans

934. Abrams, Douglas Carl. "Irony of Reform: North Carolina Blacks and the New Deal." *North Carolina Historical Review* 66:2 (1989), 149-78.

935. Bickley, Ancella. "Camp War: Remembering CCC Company 3538-C." *Goldenseal* 27 (winter 2001), 22-9.

936. Brown, E.G. "First Conference of Colored CCC Advisers." *Southern Workman* 64 (October 1935), 296-9.

937. Brown, Edgar C. *The Civilian Conservation and Colored Youth.* Washington, DC: Civilian Conservation Corps, 1939, (Mimeographed).

938. "C.C.C. Work for Negro Youth." *Monthly Labor Review* 48 (April 1939), 846-7.

939. Carvalho, Joseph. "Race, Relief and Politics: The Civilian Conservation Corps in Virginia." Master's thesis, College of William and Mary, 1977.

940. Cole, Olen, Jr. *The African American Experience in the Civilian Conservation Corps.* Gainesville, FL: University Press of Florida, 1999.

941. _____. "African-American Youth in the Program of the Civilian Conservation Corps in California, 1933-1942: An Ambivalent Legacy." *Forest & Conservation History* 35:3 (1991), 121-7.

942. _____. "Black Youth in the Program of the Civilian Conservation Corps for California, 1933-1942." Ph.D. diss., University of North Carolina at Chapel Hill, 1986.

Among the black history of the CCC in California, Cole gives the company histories and work projects performed by black CCC companies.

943. Cornebise, Alfred E. *The CCC Chronicles: Camp Newspapers of the Civilian Conservation Corps, 1933-1942.* Jefferson, NC: McFarland % Co., 2004.

A chapter of this book is dedicated to the black experience with camp newspapers.

944. Davis, John P. "A Black Inventory of the New Deal." *The Crisis* 5 (May 1935), 141-154.

945. Fishel, Leslie. "The Negro in the New Deal Era." *Wisconsin Magazine of History* 43 (Winter 1964), 111-126.

946. Gower, Calvin W. "The Struggle of Blacks for Leadership Positions in the Civilian Conservation Corps." *Journal of Negro History* 61 (April 1976), 123-135.

Gower presents a realistic picture of segregation and discrimination in the CCC by the New Deal administration and the army. He tells how Blacks were suppressed from leadership positions in the camps, and interestingly enough, how only one Black camp located at Gettysburg, PA, was a "display piece" with all Black leadership. He continues that in 1940, there were only two Black company commanders in the CCC.

947. Hoak, Michael Shane. *The Men in Green: African Americans and the Civilian Conservation Corps, 1933-1942.* Master's thesis, College of William and Mary, 2002.

948. Holland, Reid A. "Life in Oklahoma's Civilian Conservation Corps." *Chronicles of Oklahoma* 48 (Summer 1970), 224-34.

Very good article that tells a brief history of the CCC camps in Oklahoma. An interesting point in the article tells of an integrated company at Mohawk Park in Tulsa, OK, comprised of black and white veterans. Separate facilities were maintained for each race.

949. Holmes, Michael S. "The New Deal and Georgia's Black Youth." *Journal of Southern History* 38 (August 1972), 443-60.

950. Johnson, Charles W. "The Army and the Civilian Conservation Corps, 1933-1942." *Prologue 4* (Fall 1972), 139-55.

951. _____. "The Army, the Negro and the Civilian Conservation Corps: 1933-1942." *Military Affairs* 3 (October 1972), 82-87.

952. Johnson, Jesse J. *Court-Martial in the Shadow of the Trees: Blacks in the Civilian Conservation Corps, 1934; Time 1934, Location Pennsylvania.* Hampton, VA: J.J. Johnson, 1994.

953. Kifer, Allen. "The Negro and the New Deal." Ph.D. diss., University of Wisconsin, 1961.

954. Kirby, John B. *Black Americans in the Roosevelt Era: Liberalism and Race.* Knoxville, TN: The University of Tennessee Press, 1980.

955. Lapp, Rudolph. *Afro-Americans in California.* San Francisco, CA: Boyd and Fraser Publishing Co., 1979.

956. Leffler, Susan. "Nothing But Just Fighting: The 1936 CCC Race Riot." *Goldenseal* 27 (Winter 2001), 30-3

This article is based on an oral interview with Bill Dougherty, a black enrollee at Camp Loring near Minnehaha Springs, WV. Dougherty tells of a "race riot" which broke out over Black enrollees refusing to work because of severe weather conditions, and repercussions by white officers.

957. Martschukat, Jurgen and Olaf Stieglitz. "Diskurs, Erfahrung und junge African-Americans im Civilian Conservation Corps, 1933-1942." *Manner* 1 (2001).

958. Maskin, Melvin R. "Black Education and the New Deal: The Urban Experience." Ph.D. diss., New York University, 1973.

959. NAACP. *NAACP Legal File: Cases Supported, CCC Boys, 1938-1939.* Frederick, MD: University Publications of America, 1988.

This entry covers correspondence, memoranda and press releases concerning the cases of three Black CCC enrollees charges with murder in Georgia.

960. Oxley, Howard W. "The Civilian Conservation Corps and the Education of the Negro." *Journal of Negro Education* 7 (July 1938), 375-82.

961. _____. "Meeting Problems of Negro Enrollees." *School Life* 22 (January 1937), 144+.

962. Potter, Barrett George. "Civilian Conservation Corps and New York's 'Negro Question': A Case Study in Federal-State Relations During the Great Depression." *Afro-Americans in New York Life and History* 1 (July 1977) 183-200.

963. Salmond, John A. "The Civilian Conservation Corps and the Negro." *Journal of American History* 52 (June 1965), 75-88.

964. Schmitzer, Jeanne Cannella. "CCC Camp 510: Black Participation in the Creation of Mammoth Cave National Park." *Register of the Kentucky Historical Society* 93 (Autumn 1995), 446-64.

965. Wandall, Luther C. "A Negro in the CCC." *Crisis* 42 (August 1935), 244-254.

This is a first hand evaluation of life in the CCC by a black enrollee from New York. His first impressions in a "Jim Crow" atmosphere with prejudices later changes to a positive outlook for a person who has no work.

966. Wimbush, Solomon Mitchell. "Louisiana, the Civilian Conservation Corps Black Camps in District 'E': 1933-1942." Master's thesis, Louisiana State University and Agricultural and Mechanics College, 1972.

967. Wright, M.T. "Negro Youth and the Federal Emergency Programs: CCC and NYA." *Journal of Negro Education* 9 (July 1940), 397-407.

Native Americans

968. "C.C.C. Activities for Indians." *Monthly Labor Review* 49 (July 1939), 94-5.

969. Franco, Jere. "Beyond Reservation Boundaries: Native American Laborers in World War II." *Journal of the Southwest* 36 (Autumn 1994), 242-54.

This short article tells that 85,200 Native Americans participated in the CCC program. It also tells of their training and how it was beneficial to them and the country during World War II.

970. Gower, Calvin W. "The CCC Indian Division: Aid for Depressed Americans." *Minnesota History* 43 (Spring 1972), 3-13.

971. Hanneman, Carolyn G. "Baffles, Bridges, and Bermuda: Oklahoma Indians and the Civilian Conservation Corps–Indian Division." *Chroni cles of Oklahoma* 77:4 (Winter, 1999), 428.

972. Heath, V. and John Clark Hunt. "Alaska CCC Days." *Alaska Journal* 2 (Spring 1972), 51-6.

973. "The Indian C.C.C." *American Forests* 41 (September 1935), 506.

974. "Indian New Deal." *Saturday Evening Post* 211 (April 1, 1939), 22.

975. "Indians in C.C.C. Camps." *Missionary Review of the World* 56 (December 1933), 611.

976. Jonaitis, Aldona. "Totem Poles and the Indian New Deal." *Canadian Journal of Native Studies* (Canada) 9:2 (1989), 237-52.

Jonaitis discusses the government sponsored totem pole program that began in 1938. Funds were provided by the Indian CCC to sponsor Native craftsmen. Totem poles were to be restored, copied or recarved if necessary. However, the lack of training in the traditional styles affected the project. The project resulted in 48 restored poles, 54 copies and 19 new ones.

977. Killilae, Walter. "The Civilian Conservation Corps and the Sioux: The Impact of One New Deal Agency on Native Americans in the Dakotas and Minnesota." Master's thesis, University of Missouri-Kansas City, 1981.

978. LaRose, Lamar. *Life on the Reservation.* Logan, UT: Utah State University, 1973.

979. Lundblad, Dean H. "Problems in Indian Education: Past and Present. The Indian Division of the Civilian Conservation Corps [Two Plan B Reports]." Master's thesis, Utah State University, 1968.

980. Montoya, Maria E. "The Roots of Economic and Ethnic Divisions in Northern New Mexico: The Case of the Civilian Conservation Corps," *Western Historical Quarterly* (Spring 1995), 14-34.

981. Parman, Donald L. "The Indian and the Civilian Conservation Corps." *Pacific Historical Review* 40 (February 1971), 39-56.

982. _____. "The Indian Civilian Conservation Corps." Ph.D. diss., University of Oklahoma, 1967.

Known first as the IECW (Indian Emergency Conservation Work), and later the CCC Indian Division, Parman tells how the CCC contributed to the general well-being of Native American enrollees. He also tells of their work projects and how they varied because of geographical location throughout the country. Appendix II provides a summary of permanent improvements made on reservations under the CCC.

983. Parman, Donald L. and Thomas R. Wessel. "New Deal Indian Agricultural Policy and the Environment: The Papagos as a Case Study." *Agricultural History* 66:2 (1992), 22-33+.

984. Rakestraw, Lawrence. *A History of the United States Forest Service in Alaska.* Anchorage, AK: USDA Forest Service, 1994.

Although a Forest Service history, there are several pages dedicated to the work projects of the CCC in Alaska including totem pole rehabilitation and archaeological excavations at for Old Sitka.

985. Schroeder, Louis C. "Indian Conservation Camps." *Recreation*, 28 (August 1934), 249-52.

986. Sergeant, Elizabeth S. "New Deal for the Indian." *New Republic*, 95 (June 15, 1938), 151-4.

987. Sorensen, Conner. *Civilian Conservation Corps in Southeast Alaskan Newspapers: An Index.* Anchorage, AK: Alaska Historical Commission, 1985.

988. Sorensen, Conner, Lawrence Rakestraw and Robert R. Martin, Sr. *Alaskan Native Participation in the Civilian Conservation Corps: A Finding Aid to the Sources. n.p.:* USDA Forest Service, Alaska Historical Commission Studies in History No. 206, 1986

This is a very useful index to Alaskan CCC collections in Alaska and the Pacific northwest. An extensive guide to Native American (Alaskan) holdings in the National Archives is also included.

989. "Three C's in Hawaii." *American Forests* 41 (January 1935), 45.

EDUCATION & TRAINING

990. "Academic Freedom in the Civilian Conservation Corps." *School Review* 43 (January 1935), 3-7.

991. "Activities of C.C.C. in 1938." *Monthly Labor Review* 48 (February 1939), 338-40.

992. American Council on Education. American Youth Commission. *The Civilian Conservation Corps.* Washington, DC: American Youth Commission, 1940.

993. _____. *Youth in the CCC.* Washington, DC: Government Printing Office, 1942.

994. "American Youth Commission Evaluates the C.C.C." *Journal of Forestry* 39 (March 1941), 269-70.

995. Anderson, Sherwood. "Tough Babes in the Woods. Making A New Kind of American Man Out of the City Boy and Making A New Land with His Aid." *Today* 1 (February 1934), 6-7+.

996. "Appraising the CCC." *American Forests* 44 (February 1938), 75.

997. "Appraising the Civilian Conservation Corps." *School Review* 43 (September 1935), 485-9.

998. "At the Observation Post: Making Idle Boys Into Self-Reliant Youths." *Literary Digest* 117 (April 28, 1934), 12.

999. Atkinson, C. "If Opportunity Only Knocks." *Journal of Adult Education* 13 (January 1941), 59-61.

1000. Aydelott, Clarence Riley. "Facts Concerning Enrollees, Advisers, and Educational Programs in CCC Camps in Missouri." Ph.D. diss., University of Missouri, 1936.

1001. Bauder, C.F. "Vocational Education and the New Deal." *Industrial Education Magazine* 27 (May 1936), 135-6.

1002. Baurer, H.T. "Some Effects of Short-Term Trade Experience Classes Upon the Stenquist Mechanical Aptitude Test." *Education* 61 (October 1940), 97-100.

1003. Beam, Clyde C. "Civilian Conservation Corps Camps, Safety in Reclamation." *Reclamation Era* 27 (May 1937), 116-7+.

1004. Beam, Russell A. "Young Men--Old Church." *Religious Education* 33 (January 1938), 9-13.

1005. "Bee Culture in the CCC." *American Bee Journal* 80 (July 1940), 332-3.

1006. "Beekeeping in the CCC." *American Bee Journal* 80 (September 1940), 422.

1007. "Benefits Received." *School Life* 22 (June 1937), 300.

1008. Benson, R.R. "Educational Progress in CCC Camps." *Sierra Education News* 31 (January 1935), 42-4.

1009. Bimson, O.H. "Support and Control of Public Education." *Nebraska Education Journal* 22 (January 1942), 15.

1010. Blake, W.D. "NYA, the CCC and the Junior College." *School and Society* 56 (July 11, 1942), 46-7.

1011. Blakeslee, Kenneth Robert. "The Educational Program in the Fresno District, Ninth Corps Area, Civilian Conservation Corps." Ph.D. diss., Leland Stanford Junior University, 1939.

1012. Brewer, J.M. "What the CCC Might have Been." *Junior-Senior High School Clearing House* 10 (November 1935), 159-164.

1013. Brockman, C. Frank. "Park Naturalists and the Evolution of National Park Service Interpretation Through World War II." *Journal of Forest History* 22 (January 1978), 24-43.

1014. Brooks, G.H. "Education in the CCC." *Today* 4 (August 17, 1935), 18+.

1015. Brown, George Sterling. "The Educational Program of the Civilian Conservation Corps." Master's thesis, University of Oklahoma, 1935.

1016. Brubaker, A.R. "Formula for Citizenship." *Social Studies* 29 (December 1938), 339-42.

1017. Bryant, Martel Prideaux. "Education in the CCC Camps in New Mexico." Master's thesis, Texas Technological College, 1940.

1018. Butterweck, J.S. "Youth and National Needs: N.Y.A. and C.C.C. Upheld." *Nation's Schools* 29 ((March 1942), 55-6.

1019. Byrd, Thomas Smith. "An Educational Program, Through Guidance, for Youths in the Civilian Conservation Corps." Master's thesis, University of Wyoming, 1939.

1020. "CCC Camps and the American Association of University Professors." *School and Society* 40 (December 8, 1934), 766-7.

1021. "CCC Classes." *Survey* 71 (November 1935), 337.

1022. "C.C.C. Educational Advisers Named." *American Forests* 40 (March 1934), 124.

1023. "CCC Educational Program." *School and Society* 41 (April 27, 1935), 566.

1024. "CCC Illiterates." *Survey* 72 (January 1936), 20.

1025. "CCC is Builder of Youth's Morale." *Washington Post* (August 19, 1934), 12.

1026. "C.C.C. Student Enrollment." *American Forests* 41 (May 1935), 242.

1027. Caldwell, John Carroll. "A Curriculum Proposal for a Civilian Conservation Corps Camp in Rural Maine." University of Maine, 1937.

1028. Campbell, C.G. "Science Education in the CCC." *Education* 56 (March 1936), 408-9.

1029. Caravati, H.L. "Civilian Conservation Corps in Education-Recreation Field." *Recreation* 34 (February 1941), 650-1.

1030. Carr, W.G. "Educational Policy Toward Youth." *Phi Delta Kappan* 24 (February 1942), 247-8.

1031. "Citizens Conservation Corps and The Pennsylvania State College." *School and Society* 40 (October 13, 1934), 485.

1032. "Civilian Bosses for the CCC Urged in Youth Experts' Study." *Newsweek* 16 (December 30, 1940), 31.

1033. "Civilian Conservation Corps." *School and Society* 38 (August 12, 1933), 207.

1034. "Civilian Conservation Corps." *School and Society* 38 (December 16, 1933), 796.

1035. "Civilian Conservation Corps." *School and Society* 44 (December 12, 1936), 772-3.

1036. "Civilian Conservation Corps: Colleges and Universities Cooperate in Educational Program." *School and Society* 44 (December 12, 1936), 772-3.

1037. "Civilian Conservation Corps in the United States: Its Educational Programme." *International Quarterly of Adult Education* 2 (May 1935), 177-92.

1038. "Civilian Conservation Corps: Opportunities for Study and Recreation." *School and Society* 38 (August 12, 1933), 207.

1039. Clarke, F.M. and J.H. Skully. "Guidance Problems in Connecticut C.C.C. Camps." *Journal of Education Research* 32 (October 1938), 131-7.

1040. Coakler, Robert W., Richard C. Kugler and Paul J. Scheips. *Army Experience in Youth Training, 1920-1955*. Washington, DC: Office of the Chief of Military History, 1964.

1041. Cook, J.R. "Problem of Youth in the CCC." *National Education Association Proceedings* (1940), 112-3.

1042. "Correspondence Courses for the CCC Camps." *School and Society* 45 (April 3, 1937), 463-4.

1043. Couch, Paul Ellsworth. "Educational Emphasis in Civilian Conservation Camps of the Seventh Corps Area." Ed.D. diss., Indiana University, 1944.

1044. Crane, A.G. "Education by Radio in CCC Camps." *Education by Radio* 5 (September 5, 1935), 41-3.

1045. Cruey, G. Wayne. "The Educational Program of the Civilian Conservation Corps: Its Provisions for Leisure Time and Vocational Guidance." Master's thesis, Bowling Green State University, 1938.

1046. Crum, Paul Milton. "Aspects of Federal Relationships to Education." Master's thesis, University of North Dakota, 1939.

1047. Curtis, H.S. "Our Peace Time Army, the Three C's." *School and Society* 41 (February 16, 1935), 230-2.

1048. Czyszczon, Gregory John. "Progressive Education in the Civilian Conservation Corps, 1933-1942." B.A. thesis, James Madison University, 1993.

1049. Dale, Kenneth I. "The Educational Program of North Dakota Companies of Civilian Conservation Corps." Master's thesis, University of North Dakota, 1936.

1050. Danner, J.W. "What CCC Taught Me." *Rotarian* 59 (July 1941), 48-50.

1051. Darby, Francis Combs. "Public School Participation in the CCC Educational Program as Applied in the Grossmont Union High School District." Master's thesis, [University of] Oregon, 1941.

1052. Dechent, C.E. "Education for CCC." *Sierra Education News* 35 (January 1939), 18.

1053. Deming, R.C. "Education in CCC Camps." *Journal of Education* 120 (November 1937), 369-71.

1054. Dent, E.C. "Meeting an Emergency with Motion Pictures." *Nation's Schools* 14 (August 1934), 56-8.

1055. Diener, Jacqulyn M. "Major Influences of Three New Deal Agencies on Education in the United States Prior to 1944." *Journal of the Alabama Academy of Science* 43:4 (1972), 248-58.

1056. Doggett, Denzil. "Engineering in Indiana's CCC Camps." *Military Engineer* 26 (January-February 1934), 56-7.

Doggett gives an informative view of the role of the supervisory personnel in Indiana CCC camps and how each fits into work projects.

1057. _____. "Engineers Play Active Part in CCC Work in Indiana." *Engineering News-Record* (May 16, 1935), 703-705.

Twenty companies have eighty engineers on their staff, eleven in headquarters planning, twenty in state forests and game preserves, twenty in soil erosion work, and twenty-seven in state parks.

1058. Douglass, A.A. "Work Experience and the Schools." *California Journal of Secondary Education* 17 (May 1942), 297-300.

1059. Douglass, Aubrey Augustus. *Resume of the Douglas [i.e., Douglass]-Allen Claremont Survey of the CCC Education Program in the March Field and Fort MacArthur Districts.* Claremont, CA: n.p., 1936.

1060. Earp, P.B. "Curriculum Content in Civilian Conservation Corps Camps in Illinois." Master's thesis, Colorado State College of Education, 1937.

1061. "Education for 300,000: Directors and Advisers Selected to Launch Uncle Sam's Largest Peacetime Training Program in 1,468 CCC Camps." *School Life* 19 (February 1934), 117+.

1062. "Education in the CCC." *School Life* 24 (December 1938), 89.

1063. "Education in CCC Camps in February 1936." *Monthly Labor Review* 43 (November 1936), 1203-5.

1064. "Education in the CCC Camps: Objectives of the Program." *School Life* 24 (January 1939), 103-9.

1065. "Education in the Civilian Conservation Corps." *School Review* 42 (December 1934), 729-31.

1066. "Education Through the CCC." *School and Society* 52 (November 9, 1940), 455-6.

1067. "Educational Activities in the CCC Camps." *Industrial Education Magazine* 37 (September 1935), 6A-7A.

1068. "Educational Appropriations for the CCC Camps." *School and Society* 42 (July 27, 1935), 120.

1069. Educational Policies Commission. *The Civilian Conservation Corps, The National Youth Administration, and the Public Schools*. Washington, DC: Educational Policies Commission, National Education Association of the United States, and the American Association of School Administrators, 1941.

1070. "Educational Program for C.C.C. Gets Under Way: Forestry First in Fechner Outline." *Forestry News Digest* (December 1933), 1+.

1071. "Educational Program of the Civilian Conservation Corps." *Monthly Labor Review* 39 (August 1934), 375-7.

1072. "Eight Years of CCC: Training for 2,500,000 Youths Has Cost U.S. $2,779,700,000. *Newsweek* 17 (April 14, 1941), 23.

1073. Elliott, Charles Newton. "Forest Scholars." *American Forests* 41 (May 1935), 226+.

1074. "Emergency Conservation Work." *Library Journal* 59 (February 1, 1934), 120-1.

This article gives a brief description of the organizational structure of the CCC. It also tells the camp educational adviser's duties and the goals of the educational program.

1075. "Emergency Conservation Work." *Science* 79 (January 12, 1934), 27.

1076. Emerson, L.A. "Vocational Education Through Non-School Governmental Agencies." *National Society for the Study of Education Yearbook* (42nd, pt. 1), 297-317.

1077. "England to Train Jobless Youth." *School Life* 20 (November 1934), 56+.

1078. Farr, H.L. "CCC: A Rich Field for Visual Education." *Educational Screen* 13 (November 1934), 236.

1079. _____. "Civilian Conservation Corps." *School and Society* 40 (September 22, 1934), 386-8.

1080. _____. "One Teacher's View of the C.C.C." *School Life* 20 (May 1935), 211.

1081. Fear, R.A. "CCC Rebuilds Maladjusted Youth." *Phi Delta Kappan* 22 (April 1940), 385-9.

1082. Fechner, Robert. "The CCC and the Employment Service." *Employment Service News* 6 (May 1939), 3.

1083. _____. "Study Hour in the C.C.C." *Empire Forester* (1934), 33-5.

1084. "Fewer Youth Props: Plan to Merge CCC and NYA." *Newsweek* 18 (November 10, 1941), 24.

1085. Fillman, Mariette. "The Civilian Conservation Corps Model of Comprehensive Adult Education." Master's thesis, San Francisco State University, 1993.

1086. "Films in the CCC Camps: Interview with H.W. Oxley and H.H. Reimund." *Scholastic* 36 (April 29, 1940), 8T-9T+.

1087. Finch, Margaret. "Educational Opportunity in the Civilian Conservation Corps." Master's thesis, DePaul University, 1936.

1088. Finley, G.J. "Experiment in Agricultural Training." *School Life* 27 (November 1941), 52-4.

1089. Fleming, P. "University High School Boys in the Civilian Conservation Corps." *University High School Journal* 15 (October 1936), 56-61.

1090. Fox, J.A. "Challenge to CCC." *Education* 61 (October 1940), 65-8.

1091. _____. "Guiding the Forty-Year Olds." *Occupations* 17 (March 1939), 512-15.

1092. Fuller, H.S., and Johnson Hagood, Jr. *Handbook for Drivers, C.C.C.* San Antonio, TX: Pabst Engraving Co., 1935.

1093. Gage, P.S. "Three R's and the C.C.C." *American Forests* 40 (March 1934), 112-14+.

1094. Gallogly, Lester Harold. "A Study of the Personnel and Educational Program of Three Civilian Conservation Corps Camps." Master's thesis, Ohio State University, 1936.

1095. Gaumnitz, W.H. "Supervision by State Departments of Education of the Educational Activities of the Civilian Conservation Corps." *U.S. Office of Education Bulletin* 12 (1940), 57-60.

1096. "A General Tests the CCC Boy." *Literary Digest* 121 (January 25, 1936), 17.

1097. Golze, Alfred R. "Reclamation Trains the Civilian Conservation Corps Enrollees." *Reclamation Era* 29 (March 1939), 62-4.

Golze, a supervising engineer with the Bureau of Reclamation, discusses job training on reclamation projects during the years 1936-7. He tells of almost one thousand enrollees who left the Corps to obtain jobs in the work force, and the types of employment they chose.

1098. Good, Albert H. *Park and Recreation Structures: Part I-Administration and Basic Service Structures.* Washington, DC: National Park Service, 1938.

This, and the two following books are among the most important books to be written by the CCC and National Park Service in the 1930s. They contain photographs and drawings of many different types of structures used in various parts of the country by the NPS. It defines different types of architecture used in different parts of the country. They were reprinted into one complete volume in 1999.

1099. _____. *Part II-Recreational and Cultural Facilities.* Washington, DC: National Park Service, 1938.

1100. _____. *Part III-Overnight and Organized Camp Facilities.* Washington, DC: National Park Service, 1938.

1101. Gower, Calvin W. "The Civilian Conservation Corps and American Education: Threat to Local Control?" *History of Education Quarterly* 7 Spring 1967).

1102. "Great Opportunity in CCC Camps." *Missionary Review* 59 (May 1936), 255.

1103. Green, W. "Labor and the Youth Situation." *Education* 61 (October 1940), 68-70.

1104. Greene, John Thomas. "An Investigation of Certain Civilian Conservation Corps Camps to Ascertain their Educational and Vocational Needs." Master's thesis, Alabama Polytechnic Institute, 1935.

1105. _____. "Educational and Vocational Purposes of Boys in Civilian Conservation Corps Camps and Facilities for Training." Master's thesis, Alabama Polytechnic Institute, 1945.

1106. Griffing, J.B. "Educational Process in the CCC." *Sociology and Social Research* 19 (March 1935), 376-80.

1107. Harby, Samuel Farkas. *A Study of Education in the Civilian Conservation Corps Camps in the Second Corps Area, April 1933-March 1937*. Ann Arbor, MI: Edwards Brothers, Inc., 1938.

1108. _____. "A Study of Education in the CCC Camps of the 2nd Corps Area, April 1933 - March 1937." Ph.D. diss., Columbia University, 1938 (1939).

1109. Hart, James W. *Practical Construction Methods for the C.C.C. Enrollee.* Ithaca, NY: Cayuga Press, 1936.

Hart, a foreman with CCC Co. 1247, provides a reference on standard construction methods. He states that this book could be used with the on-the-job training the enrollee received, because it explained construction techniques in plain language and with drawings.

1110. Havighurst, R.J. "Youth Service in the U.S.A." *Journal of Education* (London) 74 (September 1942), 393-4.

1111. Heep, R.H. "The Civilian Conservation Corps, A New Kind of Educational and Vocational Training." Ph.D. diss., College of the City of New York, 1939.

1112. _____. "New Kind of Educational and Vocational Training." *School and Society* 50 (September 23, 1939), 410-13.

1113. Henry, Lyle K. "Civilian Conservation Corps as an Educational Institution." *School and Society* 43 (January 11, 1936), 62-6.

1114. _____. "Evaluation of the Camp Program and its Relationship to Public Education." *Internationale Zeitschrift fur Erziehung* 6 (1937), 278-9.

1115. _____. "Performance of Adults of Various Grade Levels In Specific Mental Functions: A Grade-by-Grade Comparison With School Pupils." *Journal of Education Research* 33 (October 1939), 93-101.

1116. Henthorn, B.H. Criticism of the Educational Work of CCC and NYA." *Congressional Digest* 21 (June 1942), 173-7.

1117. "Here are Educational Advisers for the C.C.C. Camps." *Forestry News Digest* (April 1934), 6-7.

1118. Herlihy, John Gilbert. "A Comparison of the Educational Programs of Two Federal Youth Programs: The Civilian Conservation Corps and the

Job Corps of the Economic Opportunity Act of 1964." Ph.D. diss., University of Connecticut, 1966.

1119. Herring, J.W. "Youth Services of the New Deal and Future Action." *Social Education* 7 (February 1943), 51-7.

1120. Hicker, Harry DeVere. *Program of Educational Activities for Emergency Camps and Shelters in California.* [Sacramento, CA]: California Department of Education Bulletin no. 5, 1934.

1121. Hicks, Thomas. "The Historical Development of the Educational Policy of the Edison Job Corps Center." Ed.D. diss., Rutgers University, 1983.

1122. Hill, Frank Ernest. "Does Youth Want to Learn?" *Journal of Adult Education* 7 (April 1935), 135-41.

1123. _____. *The School in the Camps: The Educational Program of the Civilian Conservation Corps.* New York: American Association for Adult Education, 1935.

Hill tells of the educational program of the CCC from all perspectives. He visited many of the camps and tells how innovative instructors use audio visual equipment, radios, books and other means to teach the enrollees.

1124. _____. "Vocational Guidance in the CCC." *Occupations* 14 (April 1936), 619-24.

1125. Hitchcock, A.A. "Education in the Civilian Conservation Corps." *Harvard Teachers Record* 5 (February 1935), 38-40.

1126. _____. "What C.C.C. Men Think About the Social Studies." *Social Studies* 27 (January 1936), 18-21.

1127. Holbrook, Stewart. "Forest Architecture--Old and New." *American Forests* 42 (December 1936), 558-61.

Holbrook tells how the CCC played an important part in the construction of new buildings in the national forests. He says that only with the army of youth and experienced architect supervisors the "new" building styles were adapted to the forests in which they were constructed.

1128. Holland, Kenneth. "Arts and Crafts in the CCC." *School Life* 21 (March 1936), 186+.

1129. _____. "CCC Camps and their Relation to Youth." *High School Journal* 23 (March 1940), 110-12.

1130. _____. "Implications of Work Camps for Secondary Education." *Bulletin of the National Association of Secondary School Principals* 24 (April 1940), 29-42.

1131. Holland, Kenneth and Frank Ernest Hill. *School in the Camps*. Washington, DC: American Council on Education, 1941.

1132. Holloway, Wilfred B. "Youth Employment Education Programs: Where are we Headed?" *Education and Urban Society* 14:1 (1981), 33-54.

1133. House, Darrell Clore. "The Effect of a Civilian Conservation Corps Camp's Recreational Program on the Realization of its Educational Objectives." Master's thesis, University of Kentucky, 1937.

1134. Howell, Charles E. "Federal Policies Toward Adult Education 1926-1936." Ph.D. diss., University of Iowa, 1937.

1135. "Implementation Program to Follow CCC Educational Survey." *California Schools* 10 (July 1939), 179-80.

1136. "Individual Guidance in a CCC Camp: Its Effect Upon Participation and Quality of Work in a Voluntary Education Program." *U.S. Office of Education Bulletin* 1939:7, 1-43.

1137. Irwin, William Grover. "The Work of the Chaplain in the Civilian Conservation Corps." B.D. thesis, Emory University, 1940.

1138. "Is the C.C.C. Educational Program a Success?" *School Executives Magazine* 53 (July 1934), 340-1.

1139. John, W.C. "Schools Under the Federal Government." *School Life* 26 (January 1941), 116-7.

1140. Johnson, Jed. "Educational Program of CCC Camps." *Journal of the National Education Association* 24 (February 1935), 54.

1141. Jones, Gordon Andrew. "The Teaching of Illiterates in the Civilian Conservation Corps." Master's thesis, Arizona State University, 1940.

1142. Judd, Charles H. "Educational Activities in the CCC." *The Phi Delta Kappan* 19 (May, 1937), 300.

1143. _____. "Educational Program of CCC Camps." *Journal of the National Education Association* 24 (February 1935), 54.

1144. _____. "Federal Relations With Secondary School Administrators." *North Central Association Quarterly* 15 (January 1941), 312-29.

1145. _____. "Induction of Young People Into Adulthood." *School Review* 48 (March 1940), 181-92.

1146. _____. "Real Youth Problem; Discussion of the Educational Policies Commission Pamphlet, Civilian Conservation Corps, National Youth Administration, and the Public Schools." *School and Society* 55 (January 10, 1942), 29-33.

1147. _____. "Youth Problem: Federal Services and the Schools." *Vital Speeches* 6 (April 15, 1940), 400-4.

1148. Kennedy, T.A. "Opportunity for Industrial Training in the Civilian Conservation Corps Camp Program." Master's thesis, Oklahoma Agricultural and Mechanical College, 1935.

1149. Korstain, Clarence F. *Forestry on Private Lands in the United States.* Durham, NC: Duke University Press, 1944.

1150. Kylie, Harry Raymond, G.H. Heironymus, and A.G. Hall. *CCC Forestry.* Washington, DC: Government Printing Office, 1937.

CCC Forestry is one of the best books produced by the U.S. Forest Service. It explains the life of the forest in terms and drawings that a layman can understand. It also presents simplified construction techniques for trails, bridges, towers, etc. This book is still sought after today.

1151. Lanigan, J.S. "Education in the CCC: Weapon or Feint." *Education* 61 (October 1940), 91-5.

1152. Lanpher, Harry C. "Civilian Conservation Corps: Some Aspects of Its Social Program for Unemployed Youth." *Social Service Review* 10 (1936), 623-36.

1153. Lauri, W.A. "Training in Citizenship." *Sierra Education News* 35 (March 1939), 26-7.

1154. Leach, H.G. "Education for Patriotism: Reply to C.A. Edson." *Forum* 93 (May 1935), 257-8.

1155. _____. "Soldiers as Educators: Reply to W. Davenport." *Forum* 98 (September 1937), 97-8.

1156. Leith, J.B. "Education in CCC." *Sierra Education News* 36 (October 1940), 38-9.

1157. Little, Malcolm G. "Address [Civilian Conservation Corps]." *National University Extension Association Proceedings* 17 (1934), 145-50.

1158. Lloyd, John H. "University of the Woods." *School Life* 19 (March 1934), 133.

1159. Lukens, F.E. "Are the CCC Camps Meeting the Problems of Youth?" *Proceedings of the National Education Association* (1941), 113.

1160. _____. "CCC and the Schools." *Education* 61 (October 1940), 2-7.

1161. McCormick, R. "If He Could Only Cook! CCC Cooking Class at the New Cumberland School." *Collier's* 105 (June 29, 1940), 12+.

1162. McEntee, James J. "CCC Program for Youth." *American Teacher* 24 (April 1940), 39-43.

1163. _____. "The CCC Today." *School Life* 27 (October 1941), 19-20+.

1164. McKinney, Guy D. "'Three R's' for Foresters." *Sunday Star*, Washington, D.C. (December 24, 1933), pt. 2, 3.

1165. Manning, S.C. "CCC Networks." *Radio News* 27 (January 1942), 68-70.

1166. _____. "Radio in the C.C.C." *Radio News* 17 (September 1935), 137-8+.

1167. Marchmount, Margaret. "The Human Side of the Triple C." *Minnesota Conservationist* 20 (January 1935), 12-3.

1168. Marsh, Clarence S. "C.C.C. Education One Year Old." *School Life* 20 (January 1935), 116.

1169. _____. "CCC Educational Program." *Proceedings, National University Extension Association* (1935), 62-4.

1170. _____. "C.C.C. Educational Program." *School Executives Magazine* 53 (May 1934), 266-7.

1171. _____. "Education for Adults in Civilian Conservation Corps." *National Education Association Proceedings* (1934), 275.

1172. _____. "Education in the CCC." *Child Study* 12 (February 1935), 142-3.

1173. _____. "Educational Program." *School and Society* 39 (March 31, 1934), 400-5.

1174. _____. "Educational Program." *School Life* 19 (April 1934), 178-9.

1175. _____. "Educational Program in the Civilian Conservation Corps." *Association of Urban Universities* (21st Annual Meeting, 1934), 36-48.

1176. _____. "Educational Program in the Civilian Conservation Corps." *Recreation* 28 (January 1935), 476-8.

1177. _____. "Educational Program in the Civilian Conservation Corps." *School and Society* 39 (March 31 1934), 400-5.

1178. _____. "Educational Program in the Civilian Conservation Corps During the Third Enrollment Period (April 1-September 30, 1934)." *Education Record* 16 (January 1935), 3-26.

1179. _____. "Educational Program of the Civilian Conservation Corps." *Bulletin of the Department of Secondary School Principals* 50 (March 1934), 219-29.

1180. _____. "Educational Program of the Civilian Conservation Corps." *Monthly Labor Review* 39 (August 1934), 375-7.

1181. _____. "Problems of Youth as Seen in the Civilian Conservation Corps Camps." *Proceedings of the National Education Association* (1935), 99-102.

1182. _____. "To C.C.C. Educational Advisers." *School Life* 19 (May-December 1934), 202, 222; 20 (February-March 1935), 17, 32, 54, 80, 136, 163.

1183. _____. "Vocational Training in the CCC." *Proceedings of the National Education Association* (1935), 723-30.

1184. "Marsh [Clarence S.] Education Head." *Forestry News Digest* (January 1934), 11.

1185. "Marsh Heads C.C.C. Educational Program." *American Forests* 40 (February 1934), 80.

1186. "Marsh Resigns as C.C.C. Educational Director." *American Forests* 41 (March 1935), 166.

1187. Maxwell, G.L. "CCC, the NYA, and the Public Schools." *Journal of the National Education Association* 30 (December 1941), 270.

1188. _____. "Wanted: A Federal Policy for Youth Education." *Montana Education* 18 (December 1941), 3.

1189. "Meeting at New York University of Officers of the Civilian Conservation Corps." *School and Society* 41 (February 2, 1935), 144-5.

1190. Minnesota. Department of Conservation. *Improvement Work in the State Forests by the CCC's Under Supervision of the State Forestry Division.* [St. Paul, MN]: Department of Conservation, 1935.

1191. Morrell, Fred. "CCC Youth Training." *Black Hills Engineer* 26 (December 1940), 215-33.

1192. "Motion Picture Service in the CCC." *Education for Victory* 1 (March 3, 1942), 28.

1193. Murphy, Albert John. *Individual Development and Guidance in CCC Camps. I Educational Guidance, II Vocational Guidance, III Personality Guidance.* Governors Island, NY: Office of Education, Second Corps Area, 1936.

1194. Myers, G.E. "Relation of Local Schools to Youth Work Program; Counterobjections to Judd Club Report." *School Review* 50 (October 1942), 577-83.

1195. "National Library Service for the Civilian Conservation Corps." *School and Society* 47 (February 12, 1938), 205.

1196. "New Educational Program for Civilian Conservation Corps." *American Forests* 41 (November 1935), 648.

1197. Newton, Norman T. *Design on the Land: The Development of Landscape Architecture.* Cambridge, MA: Belknap Press, 1971.

Newton devotes a chapter to state parks and the CCC. He talks of the importance of the importance of Conrad Wirth's Branch of Planning and State Cooperation and the choice of Herbert Evison as supervisor of State Park Emergency Conservation Work. Also discussed was the concept of

Recreational Demonstration Areas and the Park, Parkway and Recreational Area Study Act of 1936.

1198. Newton, Roger Hale. "Developing the CCC Camp Leisure-Time Program." *School Life* 21 (May 1936), 258-9.

1199. _____. "Leisure-Time Activities of C.C.C. Enrollees." *Monthly Labor Review* 43 (July 1936), 61-2.

1200. _____. "Recreation." *School Life* 26 (February 1941), 151-2.

1201. Noble, L. "Motion Pictures in the Educational Program of the C.C.C." *Educational Screen* 13 (June 1934), 155+.

1202. Nolling, Wilhelm. "Federal Programs to Provide Work and Training Experience for Unemployed Youth: A Comparative Analysis of the Civilian Conservation Corps and National Youth Administration and Related Legislative Proposals for the 1960's." Master's thesis, University of California, 1964.

1203. Nuttall, Leonard John, Jr. "Possible Influence on the Public Schools of the CCC Educational Program." *The School Review* 43 (September 1935), 508-13.

1204. Odell, T.C. "Learning to be New." *Clearing House* 16 (October 1941), 75-6.

1205. "Organized Labor and Academic Freedom in the Civilian Conservation Corps." *Elementary School Journal* 35 (January 1935), 321-6.

1206. "Orland CCC Camp Educational Program." *Reclamation Era* 38 (October 1938), 207.

1207. "Oxley CCC Educational Director." *American Forests* 41 (July 1935), 334.

1208. Oxley, Howard W. "American Youth Learn by Doing." *Journal of the National Education Association* 28 (April 1939), 111-12.

1209. _____. "Analysis of Enrollee Personnel." *School Life* 26 (March 1941), 187-8.

1210. _____. "Basis of Curriculum Planning." *School Life* 25 (April 1940), 220-1.

1211. _____. “Building A Model Camp Program." *School Life* 22 (November 1936), 80-1.

1212. _____. "Camp Life as a Basis for Program Planning." *School Life* 26 (November 1940), 56-7.

1213. _____. "CCC Activities of Significance in Community Youth Planning." *Harvard Teachers Record* 6 (February 1936), 28-37.

1214. _____. *CCC Camp Education: Guidance and Recreational Phases.* Washington, DC: Government Printing Office, 1938.

1215. _____. "CCC Camp Education: Guidance and Recreational Phases." *U.S. Office of Education Bulletin* 19 (1937), 1-23.

1216. _____. "CCC Camp Leisure-Time Program." *School Life* 21 (May 1936), 258-9. Also see summary--*Monthly Labor Review* 43 (July 1936), 61-2.

1217. _____. "CCC Camps Train Young Farmers." *School Life* 22 (April 1937), 235+.

1218. _____. "CCC Contributes to Human Conservation." *School Life* 22 (June 1937), 310-11.

1219. _____. "CCC Education Platform for 1936-37." *School Life* 22 (September 1936), 15+.

1220. _____. "CCC Education Undergoes Analysis." *School Life* 21 (June 1936), 291+.

1221. _____. "CCC Educational Achievements." *School Life* 25 (November 1939), 59-60.

1222. _____. "CCC Educational Advisers." *School Life* 18 (January 1933), 145-155.

1223. _____. "CCC Educational Advisers." *School Life* 25 (June 1940), 283-84.

1224. _____. "CCC Educational Plans for 1940." *School Life* 25 (October 1939), 19+.

1225. _____. "CCC Enrollees Go to College." *School Life* 24 (October 1938), 25-6.

1226. _____. "CCC Expands its Schools." *New York Times* (July 28, 1935), sec. 10, 9.

1227. _____. "CCC for American Youth." *Phi Delta Kappan* 24 (February 1942), 253-4.

1228. _____. "The CCC Goes to School." *Quartermaster Review* 19 (September-October 1939), 22-24.

1229. _____. "CCC Trains Young Men." *Adult Education Bulletin* 6 (June 1942), 137-8.

1230. _____. "CCC With CCC: Colleges Pledge Cooperation." *School Life* 21 (November 1935), 62.

1231. _____. "Certification of CCC Educational Work." *School Life* 25 (July 1940), 317.

1232. _____. "The Civilian Conservation Corps and the Education of the Negro." *Journal of Negro Education* 7 (July 1938), 375-82.

1233. _____. "Colleges and CCC Camp Education." *School Life* 22 (December 1936), 106+.

1234. _____. "Cooperation of Schools, Colleges, and State Departments." *School Life* 25 (March 1940), 187+.

1235. _____. "Cooperation with the C.C.C." *School Life* 21 (October 1935), 38.

1236. _____. "Counseling and Guidance in the CCC." *School Life* 21 (March 1936), 193-4.

1237. _____. "Curriculum of the CCC." *School Life* 24 (July 1939), 303-4.

1238. _____. "Developing Agricultural Training in CCC Camps." *School Life* 23 (January 1938), 138-9.

1239. _____. "Developing the CCC Camp Leisure-Time Program." *School Life* 21 (May 1936), 258-9.

1240. _____. "Education in the 1937-38 Corps." *School Life* 24 (December 1938), 89.

1241. _____. "Education in the CCC Camps." *School Life* 24 (January 1939), 103-9.

1242. _____. "Education in the CCC Camps." *U.S. Office of Education Bulletin* (1937, 2 vol. 1, chapter 4), 28-35.

1243. _____. "Education of the War Veteran in the CCC." *School Life* 25 (January 1940), 109+.

Oxley discusses the difference between educational activities of the WW I veterans and the regular enrollees in the CCC, since they are older men. He provides statistics on areas of educational concentration by the veterans. Also included is the American Legion's resolution of cooperation with the CCC.

1244. _____. "Educational Activities in the CCC Camps." *Junior-Senior High School Clearing House* 10 (November 1935), 139-44.

1245. _____. *Educational Activities in CCC Camps*. [Washington, DC]: Office of the Director of CCC Camp Education, March, 1938.

1246. _____. "Educational Equipment and Facilities in CCC Camps." *School Life* 24 (May 1939), 250+.

1247. _____. "Educational Program of the CCC Camps." (In Smithy, W.R., ed. *Secondary Education in Virginia* 25), 60-5.

1248. _____. "From CCC Camp to Employment." *School Life* 22 (February 1937), 183-4.

1249. _____. "Graduate Students Study CCC Education." *School Life* 23 (May 1938), 321-2.

1250. _____. "Guidance Attitudes." *School Life* 25 (February 1940), 154-5.

1251. _____. "Guidance in CCC Camps." *School Life* 24 (April 1939), 213+.

1252. _____. "Illiteracy in the CCC Camps." *School and Society* 42 (December 21, 1935), 863.

1253. _____. "Individualizing Education in the CCC Camps." *School Life* 23 (December 1937), 119-20.

1254. _____. "Inducting CCC Enrollees Into Citizenship." *School Life* 23 (April 1938), 286+.

1255. _____. "Job of the Camp Adviser." *School Life* 23 (September 1937), 24-5.

1256. _____. "Job Training in the CCC." *School Life* 21 (December 1935), 93+.

1257. _____. "Leisure-Time Activities of C.C.C. Enrollees." *Monthly Labor Review* 43 (July 1936), 61-2.

1258. _____. "Lessons for Secondary Schools From the CCC Educational Program." *Secondary Education* 5 (January 1936), 3-6.

1259. _____. "Meeting Problems of Negro Enrollees." *School Life* 22 (January 1937), 144+.

1260. _____. "New School Year Opens in Camps." *School Life* 23 (October 1937), 51. (Summary Article in *Monthly Labor Review* 46 (February 1938), 388.

1261. _____. "Organizing for Education." *School Life* 22 (October 1936), 47-8.

1262. _____. "Philosophy of CCC Camp Education." *School Life* 24 (March 1939), 181+.

1263. _____. "Reaching Two Million American Youth." *Journal of the National Education Association* 27 (April 1938), 110-12.

1264. _____. "Rehabilitation of Veterans." *School Life* 26 (May 1941), 252-3.

1265. _____. "Reviewing the Educational Program." *School Life* 26 (April 1941), 218-9+.

1266. _____. "Significant Trends in CCC Education." *School Life* 23 (March 1938), 251+.

1267. _____. "State Education Departments Aid CCC." *School Life* 24 (November 1937), 73+.

1268. _____. "Summer Plans." *School Life* 22 (May 1937), 275-6.

1269. _____. "Teaching Techniques in the C.C.C." *School Life* 21 (January 1936), 112.

1270. _____. "Training for National Defense." *School Life* 26 (July 1941), 315-6.

Oxley states the objectives for national defense training in the CCC camps. He tells how living and working in CCC camps have helped condition and train men for defense related jobs.

1271. _____. "Training of Camp Educational Advisers." *School Life* 24 (February 1939), 137+.

1272. _____. "Training the Camp Adviser." *School Life* 21 (April 1936), 208+.

1273. _____. "Trends in CCC Education." *School Life* 25 (December 1939), 77+.

1274. _____. "Twenty-Six Thousand Teachers Go To School." *School Life* 24 (June 1939), 275+.

1275. _____. "University Extension in CCC Camp Education." *National University Extension Association. Proceedings* (1936), 43-9.

1276. _____. "Use of Visual and Auditory Aids." *School Life* 25 (June 1941), 282-3.

1277. _____. "Visual Aids in CCC Education." *School Life* 23 (June 1938), 362-3.

1278. _____. "Visual Aids in the CCC Educational Program." *School Life* 25 (May 1940), 245+.

1279. _____. "Work as a Basis for Occupational Training." *School Life* 26 (December 1940), 82-3.

1280. Pack, C.L. "The Forestry Primer." (Civilian Conservation Corps Edition) *American Tree Association* (1933).

1281. Parr, Lewis A. "A Personnel Study of the School Achievement, Intelligence Socio-Economic Status, and Personality Adjustment of Civilian Conservation Corps Enrollees of Northeast Texas." Master's thesis, University of Texas, 1937.

1282. Pelley, Albert Louis. "An Investigation of Educational Opportunities in the Civilian Conservation Corps for the Ninth Corps Area." Master's thesis, University of Wyoming, 1941.

1283. Persons, W.F. "Selecting 1,800,000 Young Men for the C.C.C." *Monthly Labor Review* 46 (April 1938), 846-51.

1284. Pickett, Paul C. "Contributions of J.W. Studebaker to American Education." Ph.D. diss., University of Iowa, 1967.

1285. Platz, John C. "Partners in Training and Preservation." *CRM* (Cultural Resources Management) 17:1 (1994), 1-3.

Platz discusses the need to preserve remaining CCC structures in Oregon and Washington. He continues by telling of his experience putting together a training workshop, where Depression style building techniques were used. It was a successful partnership program between the Forest Service and Park Service. The pages following this article, by Lisa Sasser, tell how to set up a preservation workshop.

1286. "President Approves Educational Program for C.C.C." *American Forests* 40 (January 1934), 35.

1287. "President Considering Educational Program for CCC." *American Forests* 39 (December 1933), 561.

1288. "Provision for Education in C.C.C. Camps in 1938." *Monthly Labor Review* 46 (February 1938), 388.

1289. Query, Leo J. "Commercial Education in the Civilian Conservation Corps." Master's thesis, Boston University, 1938.

1290. Radcliffe, Herbert E. "A Socio-Economic Interpretation of our Educational Experience in the Civilian Conservation Corps." Master's thesis, Rhode Island College of Education, 1938.

1291. Ralston, Charles Frederick. *Adult Education as A Welfare Measure During the Great Depression: A Historical Case Study of the Educational Program of the Civilian Conservation Corps, 1933-1942.* Ed.D. diss., Pennsylvania State University, 2000.

1292. "Reclamation Work Trains Civilian Conservation Corps Men." *Reclamation Era* 27 (February 1937), 38-9.

1293. "Reducing Illiteracy in the CCC." *School and Society* 51 (April 13, 1940), 472-3.

1294. Reeves, Floyd W. et al. *The Advisory Committee on Education.* Washington, DC: Government Printing Office, 1938.

1295. Reifel, Dorothy M. "The Federal Emergency Educational Program, 1933-1936." Master's thesis, University of Cincinnati, 1936.

1296. Reinert, Wayne Leroy. "The Educational Program of the Civilian Conservation Corps." Master's thesis, University of Chicago, 1935.

1297. "Relationship of Local Schools to Youth Work Programs; Discussion by the Judd Club of the Educational Policies Commission Pronouncement: Civilian Conservation Corps, the National Youth Administration, and the Public Schools." *School Review* 50 (February 1942), 97-106.

1298. Robbins, E.B. "Three R's in the Three C's." *Michigan Education Journal* 15 (November 1937), 122-4.

1299. Robbins, L.H. "Training Youth, the CCC Way." *New York Times Magazine*, August 11, 1940, 8-9+.

1300. Rogers, Leon W. "CCC Educational Program." *Texas Outlook* 19 (July 1935), 23-4.

1301. _____. "Education in the Civilian Conservation Corps." *National Education Association Proceedings,* 1934, 293-4.

1302. Ruch, G.M. "Mathematics in CCC Education." *Mathematics Teacher* 35 (February 1942), 68-71.

1303. Sager, H.W. "Education in Civilian Conservation Corps Camps in Kansas." *Kansas Teacher and Western School Journal* 42 (February 1936), 10-11.

1304. Sargent, S.S. "Army Tries to Teach." *Christian Century* 53 (April 15, 1936), 565-7.

1305. Schwiering, O.C. "Adding A Curriculum to C.C.C. Camps." *Curriculum Journal* 9 (November 1938), 323-5.

1306. Scully, J.H. and H.J. Mahoney. "Study of Reading Ability in CCC Camps." *Education* 61 (October 1940), 101-7.

1307. Sessions, Elden B. "Educational Work of the CCC Camps in Ohio." Ph.D. diss., Ohio State University, 1937.

1308. Severson, Robert F., Jr. "The Civilian Conservation Corps: A Workface Solution." *Research in Economic History* Supplement no. 2, 1982, 121-6.

1309. Shoemaker, Blair. "The CCC and Preparedness." *American Forests* 45 (July 1939), 338.

1310. Shorey, John Charles. "Illiteracy in the United States Particularly in the Civilian Conservation Corps, and Suggestions for its Elimination." Master's thesis, George Washington University, 1939.

1311. Silcox, F.A. *Design and Construction of Forest Service Dams for Use of the Civilian Conservation Corps.* n.p.: The [Forest] Service, 1935.

1312. Sloper, Kirkland. "Adapting Counseling to C.C.C. Needs." *Occupations* 18 (January 1940), 111-13.

1313. _____. "The Effect of Individual Guidance Upon Participation and Quality of Work in a Voluntary Civilian Conservation Corps Education Program." Master's thesis, North Adams State Teachers College, 1938.

1314. _____. "Guidance Possibilities of the CCC." *Education* 61 (October 1940), 88-90.

1315. _____. *Individual Guidance in a C.C.C. Camp.* Washington: GPO, 1939.

1316. Sprouse, E. Frances. "Guidance, the Basic Function in the Educational Program of the Civilian Conservation Corps." Master's thesis, Southwest Texas State University, 1940.

1317. Stann, S.J. "Recreation in the CCC Camps." *Journal of Health and Physical Education* 12 (November 1941), 517+.

1318. "State Vocational Departments Aid CCC Education." *School Life* 24 (November 1938), 57.

1319. Stevens, E.J. "Part-Time School for C.C.C. Boys." *Agricultural Education Magazine* 10 (December 1937), 110.

1320. Stucky, Max T. *A Manual for CCC Newspapers.* Columbus, OH: Headquarters Fifth Corps Area, 1938.

1321. Studebaker, J.W. "Overview of the Civilian Conservation Corps." *Junior-Senior High School Clearing House* 10 (November 1933), 134-8.

1322. Stumpf, W.A. "Federal Aid for Education." Master's thesis, University of Chicago, 1935.

1323. Sullivan, Ralph W. "Effectiveness of the Job Training Program of the Civilian Conservation Corps." Master's thesis, George Washington University, 1941.

1324. _____."Job Training in the Reclamation Civilian Conservation Corps Camps." *Reclamation Era* 29 (July 1939), 184-6.

1325. "Survey of the CCC Educational Program." *School and Society* 42 (October 19, 1935), 529-30.

1326. Swain, Donald C. *Federal Conservation Policy.* Berkeley, CA: University of California, 1963.

1327. Tanner, P.F. "Youth in the C.C.C." *Catholic Education Review* 40 (September 1942), 406-11.

1328. Taylor, A.D. *Camp Stoves and Fireplaces.* Washington, DC: Government Printing Office, 1937.

Taylor, the Consulting Landscape Architect for the U.S. Forest Service in 1937, prepared this book for those involved in construction of cooking facilities. The book is heavily illustrated with right and wrong views of fireplace and stove construction. Many of the illustrations contain measurements for uniform construction.

1329. Ter Keurst, Arthur J. "The Problems of the Schools as Revealed by a Study of the Interests and Abilities of the Enrollees in the Civilian Conservation Corps." Ph.D. diss., Northwestern University, 1937.

1330. Terry, P.J. "Should the Schools Inherit the CCC and NYA?" *Clearing House* 16 (January 1942), 265-7.

1331. Thurman, Claude Harrison. "The Curricular and Instructional Program of the Junior White Civilian Conservation Corps Camps of the North and South Texas Districts." Ph.D. diss., University of Texas, 1940.

1332. Tipper, William J. "An Educational and Personnel Study of Three Civilian Conservation Corps Camps in District II, New Jersey." Master's thesis, Temple University, 1936.

1333. Todd, A.J. "Social Implications of the CCC." *Junior-Senior High School Clearing House* 10 (November 1935), 152-8.

1334. Touchet, Robert W. "New Deal Work Programs in Jefferson County, Texas: The Civilian Conservation Corps at Tyrell Park." Lamar University, 1972.

1335. Tower, D.B. "The Educational Implications of the Civilian Conservation Corps." Master's thesis, Niagara University, 1936.

1336. Trites, Hinson Leverett. *Development of the Educational Program in the Civilian Conservation Corps*. [Ames, IA]: n.p., 1940.

1337. _____. "Development of the Educational Program in the Civilian Conservation Corps." Master's thesis, Iowa State College, 1940.

1338. Turner, Ruth. "Status of Business Education in the Oklahoma Civilian Conservation Corps Camps." Master's thesis, Oklahoma Agricultural and Mechanical College, 1939.

1339. "Urges that CCC Care for All Boys Leaving School." *Science News Letter* 34 (September 3, 1938), 153.

1340. "Volcanic Editors: CCC Newsmen." *Literary Digest* 123 (June 12, 1937), 32.

1341. Waddell, P.R. "Christianity in the CCC." *Commonweal* 29 (March 3, 1939), 518-9.

1342. Walker, G.A. "Achievement Tests Aid C.C.C. Educational Advisers." *Occupations* 18 (January 1940), 262-5.

1343. _____. "Civilian Conservation Corps as a Builder of Physical Fitness in Youth." *Research Quarterly of the American Association for Health and Physical Education* 10 (May 1939), 97-101.

1344. Waller, J.E. "Selective Assignment in the CCC." *School Life* 27 (January 1942), 117-9.

1345. Wand, Dick. "Journal Scribe Writes About Visit to Two Forest Conservation Camps." *Southern Lumber Journal* 37 (September 1933), 32.

1346. Weersing, F.J. "Expanding Scope of Upper Secondary Education." *Junior College Journal* 11 (April 1941), 423-4.

1347. Westley, Harry A. "Education in the Civilian Conservation Camps of Montana." Master's thesis, Montana State University, 1936.

1348. "What Are the Facts?" *International Journal of Religious Education* 12 (October 1935), 10.

1349. Wheeler, H.N. "Forestry Education in the Camps of the Civilian Conservation Corps." *Central States Forestry Congress. Proceedings* (1934), 132-7.

1350. Wiedman, D.E. "Citizenship Through Training in the Civilian Conservation Corps." *Education* 61 (October 1940), 95-6.

1351. Wilson, Eustis D. "The Administration and Use of Visual Aids in the Civilian Conservation Corps." Master's thesis, Louisiana State University, 1938.

1352. Winthrop, Henry. "Problems and Prospects of the Intellectually Underprivileged." *Journal of Human Relations* 13:4 (1965), 505-23.

1353. "With the CCC." *School Life* 27 (February 1942), 148.

1354. "With the Civilian Conservation Corps." *American Forests* 39 (July 1933), 302-3.

1355. Zeitlin, Harry. "Federal Relations in American Education, 1933-1943: A Study of New Deal Efforts and Innovations." Ph.D. diss., Columbia University, 1958.

Note: Other references to training manuals and educational materials can be found in the Selected Government Documents section.

CAMP LIBRARIES

1356. Chamberlain, H.J. "CCC Libraries." *Library Journal* 63 (April 15, 1938), 299-302.

1357. _____. "Library Service in the Camps." *School Life* 23 (November 1937), 85+.

1358. Curran, Michael J. "Books Go to Camp Again: CCC Libraries in Massachusetts." *Library Journal* 60 (April 15, 1935), 315-18.

1359. "Library Work in the CCC." *Wilson Bulletin* 10 (June 1936), 688-9.

1360. "Status of CCC Camp Libraries." *Library Journal* 60 (November 15, 1935), 897.

This is a very well prepared article that details the make-up of the CCC camp library. It deals with type of materials in standard and traveling library collections. It also deals with questions by the American Library Association and Public Works Administration as to their participation in camp libraries.

1361. Taylor, Jed Harbottle. "How to Create Library Interest and Develop Use in a Widespread Organization: Presenting to CCC Educational Authorities a Plan for a Library Division." *Special Libraries* 29 (November 1938), 274-6.

Taylor discusses a reference service by mail to camp educational advisers and how applying library skills to the educational adviser would their objective of the CCC educational program. He tells how the library service polled the advisers and found what materials they wanted or needed and then provided them with either the materials or a subject bibliography to choose the materials from.

1362. _____. *Library Service for the Civilian Conservation Corps.* [New York]: American Association for Adult Education, 1937.

1363. _____. *Organization of Civilian Conservation Corps Camp Library Reading Rooms: A Manual.* New York: American Association for Adult Education, 1938.

1364. Tucker, William P. "CCC Library Work." *School and Society* 43 (January 4, 1936), 18-21.

1365. _____. "What They Read: The Reading of the Civilian Conservation Camp Enrollees." *Sierra Educational News* 33 (February 1937), 17.

1366. "U.S. to Buy 86,220 Books for C.C.C. Camp Libraries." *Publishers Weekly* 125 (May 19, 1934), 1844.

WORK PROJECTS

General

1367. Abbot, H.D. "C.C.C. Operations." *Soil Conservation* 3 (March 1938), 237-8.

1368. "America's Army of Foresters." *Sunday Star*, Washington, D.C. (April 16, 1933), 1 (Magazine section).

1369. "Association Urges Study of C.C.C. Project." *American Forests* 39 (November 1933), 520.

1370. "At the Observation Post: Making Idle Boys Into Self-Reliant Youths." *Literary Digest* 117 (April 28, 1934), 12.

1371. Bailey, Reed W. "Shackling the Mountain Flood." *American Forests* 41 (March 1935), 101-4+.

1372. Baker, J.H. "Government Poison for Wild Life: Rodent Control Aided by Emergency Conservation Work." *Bird Lore* 39 (March 1937), 107-11.

1373. Baker, Richard St.B. "Forest Camps of America." *Contemporary Review* 147 (June 1935), 715-20.

1374. Baldwin, Henry I. "Emergency Conservation Work in the United States." *Forest Crusader* (April 1934), 8-10.

1375. Blewett, Thomas James. "An Ordination Study of Plant Species Ecology in the Arboretum Prairies." Ph.D. diss., University of Wisconsin-Madison, 1981.

1376. Butler, George D. *The New Play Areas: Their Design and Equipment.* New York: A.S. Barnes, 1938.

New Play Areas, produced for the National Recreation Association, provides dimensions and layouts for sport play areas. It was a useful guide for CCC crews building recreation areas.

1377. "The CCC as Fire Fighters." *American Forests* 43 (May 1937), 250.

1378. "C.C.C. Camps Operate Nurseries Effectively." *American Forests* 40 (September 1934), 416.

This is a short article on shipping seeds to states for TVA projects. It lists what each state received.

1379. "CCC Continues Planting Trees." *American Forests* 44 (October 1938), 470.

1380. "CCC Equipment Sought for Fire Emergency." *American Forests* 48 (June 1942), 278-9.

1381. "CCC Flood Project Pays." *American Forests* 44 (November 1938), 516.

1382. "C.C.C. Forest Work Starts on Second Big Year: National Forests, the Parks and State Lands Sharing in Program." *Forestry News Digest* (April 1934), 1-3.

1383. "C.C.C. Records Accomplishments." *American Forests* 40 (February 1934), 82.

1384. "CCC Set Record in Forest Protection and Improvement Work." *Commercial and Financial Chronicle* 137 (December 30, 1933), 4631-2.

1385. "CCC Spent 1,841,000 Man-Days Fighting Fires. Emergency Conservation Work Completed April, 1933 to March 31, 1935." *Forestry News Digest* (August 1935), 2-3.

1386. "CCC Spent $29,375,938 for Trucks, etc. in Six Months." *Automotive Industry* 73 (December 14, 1935), 783.

1387. "C.C.C. Work Accomplishments for First Year." *American Forests* 40 (July 1934), 322.

1388. "CCC Work Valued at $292,000,000." *American Forests* 41 (February 1935), 84.

1389. "C.C.C. Work Valued at $579,000,000." *American Forests* 42 (February 1936), 90.

This article provides a breakdown of activities and work projects for the first 30 months of the CCC.

1390. Carnes, William G. "Landscape Architecture in the National Park Service." *Landscape Architecture* (July 1951), 145-150.

1391. Casemore, Robert. "Thespians in the Forest." *Michigan History* 85 (January/February 2001), 13-18.

The enrollees of Company 1615, on the shores of Crescent Lake, near Pontiac, Michigan, produced the play *Journey's End.* The World War I story was performed at Pontiac High School in March 1934.

1392. Church, Verne V. "CCC's and Fire Fighting." *Forest History* 14 (January 1971), 12-4.

1393. "Civilian Army in the Woods." *Harpers Magazine* 168 (March 1934), 487-97.

1394. "The Civilian Conservation Corps." *American Forests* 41 (September 1935), 530-3.

1395. "Civilian Conservation Corps Demonstrates Improved Practice in Weed Control." *Reclamation Era* 27 (June 1937), 142-44.

1396. "Civilian Conservation Corps has Translated Hundreds of Thousands of Idle Boys Into Self-Reliant Youths by Year's Service in Forestry and Soil-Erosion Control Work." *Literary Digest* 7 (April 28, 1934), 12.

1397. "Civilian Conservation Corps Recruits for New Period." *American Forests* 39 (November 1933), 516.

1398. "Civilian Conservation Corps: Symposium." *Journal of Forestry* 34 (March 1936), 284-319.

1399. "Civilian Conservation Corps to Continue Another Six Months at Full Strength." *American Forests* 39 (October 1933), 464.

1400. Collingwood, G.H. "Forestry Aids the Unemployed." *American Forests* 38 (October 1932), 574-5.

1401. "Conserving Human Resources." *Survey* 71 (March 1935), 81.

1402. "Corps to Plant Half Billion Trees." *American Forests* 42 (July 1936), 325.

Information is provided as to how many trees will be planted in each state. It also tells which varieties of trees will be planted in the different regions of the country.

1403. Croker, Thomas C., Jr. "The Longleaf Pine Story." *Journal of Forest History* 23:1 (1979), 32-43.

1404. Davenport, Walter. "Boys in the Woods." *Collier's* 99 (May 22, 1937), 12-13+.

1405. _____. "Thy Woods and Templed Hills." *Collier's* 92 (September 9, 1933), 10-12+.

1406. Dodge, N.N. "Enormous Potential Outlet for Honey." *Gleanings in Bee Culture* 62 (February 1934), 76-9.

1407. "Dramatic Broadcast by Corps at 59th Annual Meeting of the American Forestry Association." *American Forests* 40 (December 1934), 576.

1408. Dubreuil, Henri. "Forest Camps for Unemployed Young Men in the United States." *International Labour Review* (Geneva) 29 (February 1934), 199-205.

1409. Edwards, Russell T. "CCC Adds Millions to Forest Wealth of United States. " *Forestry News Digest* (January 1935), 1-2.

1410. Egan, Robert T. "Remembering the CCC: City Boys in the Woods." *American Forests* 89 (February 1983), 35-8+.

1411. _____. "Relics of the CCC." *The Conservationist* 41:3 (1986), 46-9.

1412. "Eight Years of CCC Operations, 1933-1941." *Monthly Labor Review* 52 (June 1941), 1405-13.

1413. Elliott, Charles Newton. "Forest Scholars." *American Forests* 41 (May 1935), 226+.

1414. "Emergency Conservation Work Under Reclamation Bureau." *Reclamation Era* 25 (May 1935), 92.

1415. Fechner, Robert. "Benefits of CCC Work." *Timberman* 35 (November 1933), 33-4.

1416. _____. "The CCC and Industry." *Army and Navy Journal* 75th Anniversary Special (1938), 45+.

1417. _____. "CCC and Waterfowl." *Nature Magazine* 24 (July 1934), 21-2+.

1418. _____. "The CCC in Soil Conservation." *Soil Conservation* 1 (January 1936), 1-4.

1419. _____. "The CCC Protects Nation's Forests--Nation Must Protect CCC." *Forestry News Digest* (January 1939), 24.

1420. _____. "Corps on Review." *American Forests* 41 (April 1935), 165-7.

1421. _____. "Work Ahead Justifies Permanent CCC." *American Forests* 42 (December 1936), 570.

1422. _____. "Work of the CCC." *Commonwealth* 2 (March 1935), 19-21+.

1423. _____. "A Year of C.C.C. Work." *Jeffersonian* 4 (February 1934), 3-5+.

1424. _____. "Youth and the Land." *Soil Conservation* 3 (March 1938), 236.

1425. "58 CCC Camps of Region One Issue Publications." *The Regional Review* 2:1 (January 1939), 30-1.

1426. "Fire Medals Awarded." *American Forests* 47 (February 1941), 90.

1427. Fisher, I. "Hampton's Aid to the Nation: Instruction to Men of C.C.C. Camps." *Southern Workman* 64 (April 1935), 119-20.

1428. Ford, Corey and Alastair MacBain. "Hell and High Timber." *Collier's* 103 (June 25, 1938), 12.

1429. "Forest Army Plans." *Business Week* (April 19, 1933), 15.

1430. "Forest Fire Methods for CCC." *American Forests* 44 (September 1938), 424.

1431. "Forest Work Project Speeds Forward." *American Forests* 39 (May 1933), 221.

1432. "Foresters Ask CCC be Made Permanent at 33rd Annual Meeting of Society of American Foresters." *Forestry News Digest* (January 1934), 6-7.

1433. "Forestry Employment." *Congressional Digest* 12 (April 1933), 124-5.

1434. "Governors Tell of Good Work Done by C.C.C." *Forestry News Digest* (February 1934), 6-7.

1435. Granger, C.M. "ECW and CCC." *Journal of Forestry* 31 (November 1933), 759-65.

1436. Greenleaf, Walter J. "Bloody Angle Camp." *School Life* 19 (January 1934), 94-5.

1437. Guthrie, John. "The CCC and American Conservation." *Scientific Health Monthly* 57 (1943), 401-2.

Guthrie, a former General Inspector of the CCC, tells of the accomplishments of the CCC on a nationwide basis. He concludes how the CCC help build and condition American youth and military personnel in the decade of the 1930's for World War II, although they did not know it was coming.

1438. _____. "The CCC and Forestry." *Annual Cruise* (May 1935), 30-1+.

1439. _____. "CCC as a Fire Fighting Unit." *American Forests* 45 (April 1939), 210-11+.

1440. _____. "Civilian Conservation Corps." *Military Engineer* 26 (January-February 1934), 15-19.

1441. _____. The Civilian Conservation Corps and American Conservation." *Scientific Monthly* 57 (November 1943), 401-12.

1442. _____. "Foresters, the Army, and the C.C.C." *Journal of Forestry* 32 (December 1934), 942-48.

Guthrie talks about the relationship between the foresters and army personnel. He talks about how the foresters did not reach work expectations during the year, and social differences between army and forestry personnel.

1443. _____. "Now They Are Men: The Story of the CCC (Review)." *Journal of Forestry* 39 (January 1941), 77.

1444. Handlan, John W. "A Municipal Park that Plants Trees." *American Forests* 44 (August 1938), 363-5+.

CCC Company 3529, Camp Waddington, was located at Oglebay Park, in Wheeling, WV. This was the only CCC camp located in the northern panhandle of West Virginia. Tree planting and roadside improvement were the main projects of this camp.

1445. Hawes, Austin F. "The C.C.C. as an Agency for Stand Improvement." *Journal of Forestry* 34 (July 1936), 664-70.

Hawes gives an overview of the timber stand improvement in Connecticut. He provides tables of how forestry work camp projects were distributed throughout the state and the cost of maintaining timber stand improvements for the first two and one-half years of the CCC.

1446. _____. "Looking Ahead with the C.C.C." *American Forests* 41 (May 1935), 214+.

1447. Hazard, J.O. (Hap). "The Roosevelt Forestry Song." *American Forests* 39 (June 1933), 250.

1448. Headley, Roy. "Why Big National Forrest Fires?" *American Forests* 40 (December 1934), 559-61+.

1449. Hearon, Fanning. "Recreation Renaissance: Development of State and County Parks." *Recreation* 29 (September 1935), 289-93+.

1450. Hersel, George. "Legacy of the CCC." *Wonderful West Virginia* 47 (December 1983), 3-11.

1451. Hieronymus, George H. "Make Them Conservees For Life." *Journal of Forestry* 32 (December 1934), 930-3.

The article talks about the relationship of the CCC camp to forestry in general. The author asks why the educational program stresses general education rather than forestry education.

1452. "High, Continuous Truss Bridge Erected by C.C.C. Crews." *Engineering News-Record* 113 (June 24, 1937), 947-8.

1453. Hoar, Crosby A. "The Civilian Conservation Corps." *Forestry Chron-icle* 10 (February 1934), 43-53.

Hoar gives a good overview of the organization of the CCC. He provides information on the formation of companies, selection of enrollees, wages of enrollees, company staff, foremen and superintendents. He also describes the duties of the agencies involved (lists) and the types of work projects that are authorized.

1454. House, Darrell Clore. "The Effect of a Civilian Conservation Corps Camp's Recreational Program on the Realization of its Educational Objectives." Master's thesis, University of Kentucky, 1937.

1455. Howard, William G. "The Civilian Conservation Corps and the Forests." *Bulletin to the Schools* 20 (April 2, 1934), 164-167.

1456. Hubbs, Carl L. "The C.C.C. at Work for Fish and Game." *National Waltonian* 1 (September 1933), 4-5.

1457. Humphrey, Hubert H. "Plan to Save Trees, Land, and Boys." *Harper's Magazine* 218 (January 1959), 53-7.

1458. Hyde, Gerald H. "Fire Troupers of the CCC." *The Regional Review* 2:6 (June 1939), 27-8.

1459. "Isle Royale Purchase Approved." *American Forests* 42 (January 1936), 33.

1460. Jackson, James P. "CCC Handiwork: Enduring Monuments." *American Forests* 94 (September/October 1988), 45-7.

1461. _____. "Living Legacy of the CCC." *American Forests* 94 (September/October 1988), 37-44.

1462. Janney, J. "C-Boys to the Rescue." *American Magazine* 131 (June 1941), 100-4.

1463. Jones, B.W. "The Valley of Lost Hope." *American Forests* 40 (July 1934), 298-300.

1464. Kauffman, Erle. "Civilian Foresters in Flood Time." *American Forests* 39 (October 1933), 454-5.

1465. Korstain, Clarence F., *et. al.* "The Civilian Conservation Corps." *Journal of Forestry* 34 (March 1936), 284-319.

1466. Leopold, Aldo. "Coon Valley: An Adventure in Cooperative Conservation." *American Forests* 41 (May 1935), 205-8.

1467. Lindley, Ernest K. "Civilian Conservation Corps and Other Federal Aid Projects." *School and Society* 41 (April 27, 1935), 578-80.

1468. _____. "Forest Interlude: After Their Service in the CCC is Concluded, How Shall Young Americans Find Their Jobs in our World of Today?" *Today* 2 (September 22, 1934), 14.

1469. Livermore, Norman B., Jr. "Roads Running Wild." *American Forests* 44 (April 1938), 153-5.

1470. MacArthur, Douglas. "The Army and Reforestation." *Infantry Journal* 40 (July-August 1933), 250.

1471. _____. "Civilian Conservation Camp Libraries." *Library Journal* 58 (October 1, 1933), 790-1.

1472. "Manicuring the Wilderness." *Saturday Evening Post* 207 (December 8,1934), 22.

1473. McEntee, James J. "The CCC and State Parks." *The Regional Review* 4:5 (May 1940), 35.

1474. McIntosh, Phyllis. "The Corps of Conservation." *National Parks* 75:9 (2001), 23+.

1475. McKinney, Guy D. "Strong CCC Fights Forest Blight." *New York Times* (August 25, 1935), 11-E.

1476. Morrell, Fred. "Men, Trees and Game." *American Forests* 42 (August 1936), 363-5+.

Morrell defends the practice of cutting fire breaks and roads in forests to protect against forest fires. The CCC is criticized in its conservation work as starving animals, fish and birds. One critic calls the corps the "circus of destruction."

1477. _____. "Straw Men in the Wilderness." *Nature Magazine* 29 (March 1937), 174-6.

1478. Morse, H.B. "Foresters Under Fire in E.C.W." *Journal of Forestry* 34 (April 1936), 388-90.

This article begins by saying that CCC project expenses were excessive and that foresters were going through a process of degeneration. It agrees that expenses could be excessive, but the foresters are not degenerating. Morse encourages the foresters to make the most of their CCC help. It asks the forester what he is doing to help the enrollee become a better worker and leader.

1479. "National Park Standards: A Declaration of Policy." *National Parks Bulletin* 14 (December 1938), 5-7.

1480. "National Register of Historic Places: Multiple Property Submissions." *CRM* (Cultural Resources Management) 19:9 (Supplement, 1996), 1-16.

This listing consists of all Thematic Resources (TM), Multiple Resource Area (MRA) and Multiple Property Submission (MPS) nominations submitted to the National Register of Historic Places. It includes CCC, ECW and Depression era nominations submitted through October 1996.

1481. Nelson, Bryce. "FDR's Trees." *American Forests* 88 (February 1982), 14-6+.

1482. Newton, Roger Hale. "Pattern for Parks." *Architectural Forum* (December 1936), 491-510.

1483. Otis, Alison T., William D. Honey, Thomas C. Hogg and Kimberly K. Lakin. *The Forest Service and the Civilian Conservation Corps: 1933-1942*. Washington, DC: USDA Forest Service, 1986.

1484. Ouderkirk, Nancy M. "The Civilian Conservation Corps: Roosevelt's Robin Hoods." *The Northern Logger and Timber Processor* 23 (May 1975), 22-3, 56-8.

This article gives a very good overview of the organization of the CCC. Unlike most articles, it gives examples of complaints and criticisms of the CCC program.

1485. Pack, Arthur N. "Forests and Labor." *Forestry News Digest* (June 1934), 5.

1486. Pack, C.L. "The Baptism of the Foresters." *Review of Reviews and World's Work* 90 (November 1934), 52-5.

1487. _____. "Blister Busters." *Scientific American* 150 (February 1934), 61-3.

1488. _____. "The Forestry Primer." (Civilian Conservation Corps Edition) *American Tree Association* (1933).

1489. "Park Plantings Soar." *American Forests* 43 (July 1937), 356.

1490. Peterson, Charles S. and Linda E. Speth. "Administration and Personnel on the Wasatch-Cache National Forest." *Journal of Forest History* 26:1 (1982), 4-19.

1491. Pitman, Leander. "Suppressing the Red Menace." *Overland Monthly* 91 (November 1933), 153+.

1492. Plair, T.B. "How the CCC Has Paid Off." *American Forests* 60 (February 1954), 28-30, 44-5.

1493. Porter, John A. "The Enchanted Forest." *Quartermaster Review* 13 (March-April 1934), 5-17.

1494. "President Roosevelt to Request Continuation of Civilian Conservation Corps: Fechner's Report Shows Forestry Program Greatly Advanced by Work." *American Forests* 40 (November 1934), 540.

1495. "Principles and Procedures for Selection of Men." *Monthly Labor Review* 40 (May 1935), 1162-9.

1496. Pyne, Stephen J. "Fire Policy and Fire Research in the U.S. Forest Service." *Journal of Forest History* 25 (1981), 64-77.

1497. "Rebuilding Paul Bunyan's Empire." *Popular Mechanics* 73 (May 1940), 674-7.

1498. "Recent Activities of the C.C.C." *Monthly Labor Review* 41 (November 1935), 1209-11.

1499. "Reclamation Camps Lead CCC in Safety in 1937." *Reclamation Era* 38 (September 1938), 192.

1500. "Reclamation Work Trains Civilian Conservation Corps Men." *Reclamation Era* 27 (February 1937), 38-9.

1501. "Reforestation Assumes Extensive Proportions." *American Forests* 40 (June 1934), 279.

1502. "Reforestation, At Last!" *Scientific American* 149 (July 1933), 9.

1503. "Reviewing Work of the Boys in C.C.C. Camps." *Forestry News Digest* (February 1934), 8.

1504. Ringland, Arthur C. "The Civilian Conservation Corps and the Natural Resources Board." *Journal of Forestry* 32:9 (December 1934), 949-51.

The article talks about how the Natural Resource Board and the CCC can work together to form a land use policy. It states that the political set-up of many federal conservation bureaus and state agencies had made it impossible to attack land use problems.

1505. Roland, Carol. *CCC Resources in the State Park System, An Inventory.* n.p.: The Resources Agency, 1991.

1506. Russell, Carl P. "Coordination of Conservation Programs." *The Regional Review* 6 (January-February 1941), 13-19.

1507. "Saving Forests From Extinction: A National Plan." *United States News* (June 24-July 1, 1933), 11.

1508. Shockly, Linda. "Three Billion Trees." *Blue Ridge Country* (September/October 1994), 16-19.

Provides reminiscence of a former enrollee in Virginia. Gives facts on CCC work projects.

1509. Silcox, Ferdinand A. "Men Against the Flames." *American Forests* 45 (April 1939), 194-8.

1510. _____. "Our Adventure in Conservation: The CCC." *Atlantic Monthly* 160 (December 1937), 714-22.

1511. "686,709 Man Days Fighting Forest Fires." *Forestry News Digest* (September 1934), 8.

1512. "Six Years of CCC Operations." *Monthly Labor Review* 49 (December 1939), 1409-12.

1513. Smith, Kenneth G. "At A Civilian Conservation Corps Camp." *Recreation* 23 (August 1934), 248+.

1514. Society of American Foresters. "Report on the Civilian Conservation Corps." *Journal of Forestry* 34 (March 1936), 307-19.

This report brings to light many questions regarding the CCC such as political appointments to supervisory positions on the local to the national level. It also examines the question of dual control of CCC camps by the Army and the work project agency.

1515. _____. "Statement Supplementary to the 1935 Report of the C.C.C. Committee of the Society." *Journal of Forestry* 34 (March 1936), 302-6.

This statement addresses the questions proposed in the above article.

1516. "State and Federal Agencies Join Forces in Park Programs." *State Government* 12 (September 1939), 171-4.

1517. Stepenoff, Bonnie. "The New Deal's Camp Sherwood Forest: An Incubator of Democracy." *Gateway Heritage* 11:3 (1990-91), 52-9.

1518. Stockwell, H.J. "Soil Erosion Control Work by the CCC." *Agricultural Engineering* 16 (May 1935), 194-7.

1519. Stone, Percy N. "Urges C.C.C. as Permanent Guard for United States Forests." *Evening Star*, Washington, D.C. (November 26, 1933), 3.

1520. Story, Isabelle F. *Glimpses of Our National Parks*. Washington, DC: Government Printing Office, 1936.

While this pamphlet contains a short history of many of the nation's parks, it contains illustrations of CCC enrollees performing work tasks. It also has a map locating national park projects, war memorial projects, national monument projects, battlefield national monument projects and miscellaneous projects on the inside back cover.

1521. _____. *The National Parks and Emergency Conservation Work*. Washington, DC: Government Printing Office, 1936.

1522. "The Story of State Parks." *American Forests* 41 (September 1935) 476-81.

1523. Sullivan, Ralph W. "Job Training in the Reclamation Civilian Conservation Corps Camps." *Reclamation Era* 29 (July 1939), 184-6.

1524. "Summary by States of C.C.C. Work Shows Great Advance Made in Forest Program: Much Fire Fighting." *Forestry News Digest* (November 1934), 8-12.

1525. Talbot, Hammond. "Forest Conservation--A Vast Problem Confronting the Whole Nation." *Washington Post* (May 19, 1935), 9-B.

1526. Taylor, Charles H. "CCC in the Wildlife Program." *Transactions of the 5th American Wildlife Conference* (1940), 41-46.

1527. Templeton, H.G. "Erosion Control Work Serves Nation's Youth." *Purdue Agriculturist* 28 (February 1934), 33+.

1528. Tharpe, L.E. "Conservation Camps and the War Department." *Christian Century* 50 (August 30, 1933), 1089.

1529. Thone, Frank. "Are We Creating an American Sahara?" *Sunday Star*, Washington, D.C. (August 12, 1934), Magazine 6.

1530. "Three C's in Hawaii." *American Forests* 41 (January 1935), 45.

1531. "Trailer Camps in State Parks Built by Uncle Sam's CCC." *National Petroleum News* 29 (June 2, 1937), 41-3.

This article tells of the newest style of tourist parks being constructed by the CCC. It gives brief descriptions of the trailer parks being constructed in each state, and the facilities available in them.

1532. Traphagen, Theodore B. "A CCC Fire-Fighter Speaks Out." *American Forests* 44 (December 1938), 530.

1533. Tweed, William C., Laura E. Soulliere and Henry G. Law. *National Park Service Rustic Architecture: 1916-1942*. San Francisco: National Park Service, 1977.

1534. Tyner, George P. "The Enrollee and Soil Conservation." *Soil Conservation* 3 (March 1938), 239.

1535. "Upsetting Nature's Equilibrium." *Horticulture* 16 (June 15, 1938), 264.

1536. Urbanek, John. "The Forest Corps." *American Forests* 41 (January 1935), 18-21.

1537. Vetter, Roy A. *Digest of Laws Relating to Local Parks and Recreation as of January 2, 1940*. Washington, DC: Federal Security Agency, 1940.

1538. Vivian, C.H. "300,000 Foresters Take to our Wooded Areas: Civilian Conservation Corps." *Compressed Air Magazine* 39 (January 1934), 4316-22.

1539. Vought, Robert H. "The Civilian Conservation Corps in Our State Forests." *Pennsylvania Department of Forests and Waters. Service Letter* 5 (January 18, 1934), 1-2.

1540. Wallace, Henry A. "Forestry Faces Challenge Says Wallace in his Annual Report." *Forestry News Digest* (December 1933), 1+.

1541. Watkins, Clayton W. "In the Dust Storm Country They Work on Erosion." *Forestry News Digest* (September 1934), 6.

1542. Wharton, William P. "Park Service Leader Abandons National Park

Standards." *National Parks Bulletin* 14 (June 1938), 4-6.

1543. "What the C.C.C. Accomplished: An Outline of Results Obtained by One of the First Relief-Public Works Programs Launched by President Roosevelt." *West Coast Lumberman* 61 (January 1934), 6-7+.

1544. "What Leaders Say of Forest Project." *Forestry News Digest* (June 1933), 3.

1545. Wheeler, H.N. "Forestry Education in the Camps of the Civilian Conservation Corps." *Central States Forestry Congress. Proceedings* (1934), 132-7.

1546. Wills, W.W. "C.C.C. Stone and Log Craft." *Industrial Education Magazine* 39 (November 1937), 250-3.

1547. Wirth, Conrad L. *The CCC and its Contribution to a Nation-Wide State Park Recreational Program.* Washington, DC: National Park Service, n.d.

1548. _____. "Parks for the Millions: The Nation-Wide Program for More Recreational Areas Grows Under the C.C.C." *American Forests* 42 (November 1936), 505-7+.

Wirth tells of the accomplishments of CCC workers in the areas of state and national parks, national monuments and recreational demonstration areas. He also tells of the cooperation with other federal agencies in recreational construction.

1549. "With the Civilian Conservation Corps." *American Forests* 39 (July 1933), 302-3.

1550. "With the Civilian Conservation Corps." *Recreation* 29 (August 1935), 265+.

1551. "Work Ahead Justifies Permanent CCC Says Fechner." *American Forests* 42 (December 1936), 570.

1552. "Work-Relief Program--Civilian Conservation Corps." *Congressional Record* 79 (March 14, 1935), 3752-55.

1553. "Work Relief Spending Plans." *News Week* 5 (April 20, 1935), 7.

1554. Wyckoff, N. "Accomplishments of the Civilian Conservation Corps in Blister-Rust Control." *Northwest Science* 8 (June 1934), 5-8.

1555. Young, Dwight. "The Work of the Civilian Conservation Corps Lives Around Us." *Historic Preservation: Quarterly of the National Council for Historic Sites and Buildings* 56:1 (2004), 64.

1556. Zon, Raphael. "Back of it All--Forest Research." *Forestry News Digest* (June 1934), 4.

1st Corps Area

1557. Clapp, Robert T. "The Effects of the Hurricane upon New England Forests." *Journal of Forestry* 36 (December 1938), 1177-81.

1558. Evans, R.M. "Civilian Conservation Corps Improves Stands on Eastern National Forests." *Journal of Forestry* 331 (November 1933), 771-7.

1559. Hawes, Austin F. "Emergency Conservation Work in Connecticut." *Wooden Nutmeg* 9 (September 1933), 1+.

1560. Holmes, Frederick C. "Timber-Truss Bridge and Dam Built by CCC Labor in Windsor, Mass." *Engineering News-Record* (April 9, 1936), 526.

1561. Pasco, Mary E. "Connecticut Improves Her Trout Streams: Civilian Conservation Corps Inaugurates Pioneer Work in Developing an Important Recreation Asset." *American Forests* 40 (February 1934), 73-5.

1562. Plumb, G.H. and H.W. Hicock. "Insect Control Work by the Civilian Conservation Corps in Connecticut." *Journal of Economic Entomology* 27 (April 1934), 344-5.

1563. Rich, Albert T. "The CCC in New England." *Quartermaster Review* 14 (January-February 1935), 5-13.

1564. Rutledge, R.H. "C.C.C. Work on Grazing District Lands." *National Wool Grower* 32 (October 1942), 13-5.

1565. Seaman, James A. "Enrollee Aid in Northeastern Orchards." *Soil Conservation* 3 (March 1938), 243.

1566. Stetson, Frederick W. "The Civilian Conservation Corps in Vermont." *Vermont History* 46 (Winter 1978), 24-42.

2nd Corps Area

1567. Evans, R.M. "Civilian Conservation Corps Improves Stands on Eastern National Forests." *Journal of Forestry* 331 (November 1933), 771-7.

1568. Marshall, Robert. "Facilities in Osborne's Position [re: Truck Trails]: An Open Letter to the Conservation Commissioner of New York." *The Living Wilderness* 1 (September 1935), 4-5.

1569. Torrey, Raymond H. "Truck Trails in the Adirondacks." *The Living Wilderness* 1 (September 1935), 3-5.

3rd Corps Area

1570. Clepper, Henry. "20,000 Men in Penn's Woods." *American Forests* 39 (September 1933), 406-7.

1571. Evans, R.M. "Civilian Conservation Corps Improves Stands on Eastern National Forests." *Journal of Forestry* 331 (November 1933), 771-7.

1572. Lotspeich, Stephen Henry. "The Design Intentions and the Planning Process of the Virginia CCC State Park Master Plans." Master's thesis, University of Virginia, 1984.

1573. "Pennsylvania Fishing Improved by the C.C.C." *American Forests* 41 (January 1935), 41.

1574. "The Public Works Administration Project to Restore the Chesapeake and Ohio Canal." *The Federal Architect* (October 1938), 42-3.

1575. Simmons, Dennis Elwood. "The Creation of Shenandoah National Park and the Skyline Drive, 1924-1936." Ph.D. diss., University of Virginia, 1978.

1576. Wales, H.B. "Cultural Work in the North Central States Region." *Journal of Forestry* 31 (November 1933), 766-70.

4th Corps Area

1577. Buell, Jesse H. "Results of C.C.C. Timberstand Improvement on Southern Appalachian National Forests." *Journal of Forestry* 41:2 (February 1943), 105-12.

1578. Dyer, Delce and Quentin Bass. "Southern Appalachian National Forests: Interpretive Planning for Rural Historic Landscapes." *CRM* (Cultural Resources Management) 17:7 (1994), 23-27.

1579. Guthrie, John. "CCC's Contribution to the South." *Southern Planter* 96 (July 1935), 5+.

1580. Hurst, Fred. "Big Erosion Job: Two Thousand C.C.C. Men at Work Battling Washouts in Ten Counties in Northern Mississippi." *Forestry News Digest* (December 1933), 13.

1581. "In the Tennessee Valley the C.C.C. Wars Against the Erosion Enemy." *American Forests* 40 (October 1934), 448-9.

1582. Lawliss, Lucy Ann. "The Civilian Conservation Corps and the State Park: An Approach to the Management of the Designed Historic Landscape Resources at Franklin D. Roosevelt State Park, Pine Mountain, Georgia." Master's thesis, University of Georgia, 1992.

Lawliss talks of the adaption of design where nature dominates in construction of the F.D.R. State Park. She also provides some photos and drawings which reinforce the rustic design theory.

1583. Lentz, Gus. "Making Water Walk in the Tennessee Valley." *American Forests* 40 (May 1934), 200-2+.

1584. Manchester, E.R. "Our Fugitive Farm Lands: Lets Move up to the Front Where CCC Workers in Mississippi are Using Explosives to Check Soil Erosion." *DuPont Magazine* 28 (February 1934), 1-3.

1585. "Reforestation by the Tennessee Valley Authority." *Science* 78 (December 1, 1933), 500.

1586. Shofner, Jerrell H. "Roosevelt's 'Tree Army:' The Civilian Conservation Corps in Florida." *Florida Historical Quarterly* 65 (April 1987), 433-56.

1587. Shoulders, Eugene. "Timber Stand Improvement on Ozark Forests -An Appraisal After 15 Years." *Journal of Forestry* 54 (December 1956), 824-7.

1588. Stauffer, Alvin P. and Charles W. Porter. "The National Park Service Program of Conservation for Areas and Structures of National Historical Significance." *Mississippi Valley Historical Review* 30 (June 1943), 25-48.

1589. Sullivan, John J. "The Civilian Conservation Corps and the Creation of the Myakka River State Park." *Tampa Bay History* 9 (Fall/Winter 1987), 4-16.

1590. Sumner, Jim L. "The Civilian Conservation Corps and the Development of North Carolina's State Park System." *Carolina Comments* 40 (July 1992), 98-106.

1591. Tate, Roger D., Jr. "George D. Power and New Deal Work Relief in Mississippi, 1933-1934." *Journal of Mississippi History* 46:1 (1984), 1-16.

5th Corps Area

1592. Cobb, Mary. "Corpsmen Built Museum At Hawk's Nest." *Wonderful West Virginia* 47:10 (December 1983), 24-25.

1593. Doggett, Denzil. "Engineering in Indiana's CCC Camps." *Military Engineer* 26 (January-February 1934), 56.

1594. _____. "Engineers Play Active Part in CCC Work in Indiana." *Engineering News-Record* (May 16, 1935), 703-705.

1595. Lumpee, Henry L. "The Civilian Conservation Corps: Irrigator." *Reclamation Era* 26 (November 1936), 268-9.

Lumpee talks about the Vale, Oregon Project. The CCC completed the project with construction of the Willow Creek lateral system. Many enrollees from the Fifth Corps Area worked on this project.

1596. Schmitzer, Jeanne Cannella. "CCC Camp 510: Black Participation in the Creation of Mammoth Cave National Park." *Register of the Kentucky Historical Society* 93 (Autumn 1995), 446-64.

6th Corps Area

1597. Rosentreter, Roger L. "Michigan's 83 Counties: Oscoda County." *Michigan History* 73:4 (1989), 8-10.

1598. Sorenson, Gordon. "The Civilian Conservation in Wisconsin." *Timber Producer* (March 1977), 26-9.

7th Corps Area

1599. Benson, David R. *Stories in Log and Stone: The Legacy of the New Deal in Minnesota State Parks.* St. Paul, MN: Minnesota Department of natural Resources, Division of Parks and Recreation, 2002.

A good deal of this book is dedicated to the the CCC in Minnesota. It talks about how the New Deal got organized, the work projects of the CCC, and the development of parks and park structures.

1600. Butler, Ovid. "Prairie Shelter Belt: Thirty Questions and Their Answers Concerning President Roosevelt's Latest Undertaking in Forestry." *American Forests* 40 (September 1934), 395-8.

1601. Durrell, Glenn R. "Economic Effects of the Great Plains Shelterbelt in Term of Soil and Human Betterment." *Journal of Forestry* 37 (February 1939), 144-7.

1602. Minnesota. Department of Conservation. *Improvement Work in the State Forests by the CCC's Under Supervision of the State Forestry Division.* [St. Paul, MN]: Department of Conservation, 1935.

1603. Peck, Allen S. "Our Forest Lands." *Black Hills Engineer* 26 (December 1940), 234-45.

1604. Rutledge, R.H. "C.C.C. Work on Grazing District Lands." *National Wool Grower* 32 (October 1942), 13-5.

1605. VanWest, Carroll. "'The Best Kind of Building:' The New Deal Landscape of the Northern Plains." *Great Plains Quarterly* 14:2 (1994), 129-41.

8th Corps Area

1606. Butler, Ovid. "Prairie Shelter Belt: Thirty Questions and Their Answers Concerning President Roosevelt's Latest Undertaking in Forestry." *American Forests* 40 (September 1934), 395-8.

1607. Granger, C.W. "CCC and Soil Conservation in the Southwest." *Soil Conservation* 2 (February 1937), 161-4+.

1608. Guthrie, John. "With the Texas Forest Army." *American Forests* 39 (December 1933), 347-8+.

1609. Martin, John S. "Accomplishments and Finances of Emergency Conservation Work Camps." *Reclamation Era* 25 (March 1935), 58-9.

Martin provides summaries of Bureau of Reclamation projects in Nebraska, New Mexico, Texas and Utah. He also lists camp allotments for 1935.

1610. Mealey, W.P. "Civilian Conservation Corps Erosion and Flood Control Program, Rio Grande Project." *Reclamation Era* 27 (July 1937), 164-5+.

1611. Neef, Diana L. "Civilian Conservation Corps: Elko County, 1933-1942." *Northeastern Nevada Historical Society Quarterly* (1984), 16-24.

1612. Richardson, Elmo R. "The Civilian Conservation Corps and the Origins of the New Mexico State Park System." *Natural Resources Journal* 6 (April 1966), 248-67.

1613. Rutledge, R.H. "C.C.C. Work on Grazing District Lands." *National Wool Grower* 32 (October 1942), 13-5.

1614. Sallee, Katherine Northcut. "The Cultural Landscape Inventory: A Research Process for the Platt District of the National Park Service (Oklahoma)." Master's thesis, University of Texas at Arlington, 1997.

1615. Schrems, Suzanne H. "A Lasting New Deal Legacy: The Civilian Conservation Corps, The National Park Service and the Development of the Oklahoma Park System." *Chronicles of Oklahoma* 72 (Winter 1994-95), 369-95.

1616. Seymour, Gregory R. "The Civilian Conservation Corps in Arizona: A Context for Erosion-Control Features." *The Kiva* 63 (1998).

1617. Steely, James Wright. "Rustic Style in Depression Texas: Federal Architecture in the State Parks, 1933-1941." Master's thesis, University of Texas at Austin, 1985.

1618. Toulouse, J.H. "CCC in the Sandias." *New Mexico Magazine* (September 1933).

1619. Wilson, R.E. "Civilian Conservation Corps Work in the South-Western United States." *Forestry Chronicle* 11 (December 1935), 317-22.

9th Corps Area

1620. Ashline, A.N. "Civilian Conservation Corps Enrollees Salvage Owyhee Railroad." *Reclamation Era* 29 (June 1939), 141-3.

1621. Ayres, R.W. "Fire and Water! How the C.C.C. is Solving These Acute Problems in California." *American Forests* 40 (April 1934), 153-5.

1622. Beam, Clyde C. "Accomplishments of Bureau Civilian Conservation Corps Camps." *Reclamation Era* 27 (September 1937), 222-3.

The author provides information on the accomplishments of the Bureau of Reclamation CCC camps. The article contains a list of Reclamation projects and the number of CCC camps assigned to each project.

1623. _____. "Civilian Conservation Corps Enrollees Operate Construction Equipment on Deschutes Project." *Reclamation Era* 31 (March 1941), 90-1.

1624. Beckham, Stephen Dow. *An Inventory and Evaluation of the Historical Significance of the Civilian Conservation Corps Buildings on the Siskiyou National Forest.* Portland, OR: S.D. Beckham, 1979.

1625. Berg, Irving. "Civilian Conservation Corps Constructs the Anita Dam, Huntley Project." *Reclamation Era* 28 (February 1938), 38-9.

1626. "C.C.C. Camps Assigned to Bureau of Reclamation." *Reclamation Era* 25 (June 1935), 129.

This article tells the location of 37 CCC camps assigned to the Bureau of Reclamation in 12 states.

1627. "C.C.C. Camps on Reclamation Projects." *Reclamation Era* 25 (September 1935), 188.

This article provides a list of all Bureau of Reclamation CCC camps for 1935.

1628. California. Department of Parks and Recreation. *CCC Resources in the State Park System, An Inventory* (by Carol Roland). n.p.: The Resources Agency, 1991.

1629. "Civilian Conservation Corps Constructs 18-Mile Range Drift Fence, Owyhee Project." *Reclamation Era* 29 (May 1939), 115.

1630. "Civilian Conservation Corps Program for Federal Reclamation Projects, Fiscal Year 1939." *Reclamation Era* 28 (August 1938), 163.

1631. "Civilian Conservation Corps Rebuilds Deer Flat Reservoir Embankments." *Reclamation Era* 28 (November 1938), 230-1.

1632. "Civilian Conservation Corps Work to Continue on Reclamation Projects." *Reclamation Era* 28 (April 1938), 74.

1633. Crancer, John and Bruce Reichart, *et.al. CCC in Idaho.* Boise, ID: Idaho Public Television, VHS, 1999.

1634. Dunkley, L.R. "Civilian Conservation Corps Builds Midview Reservoir and Canals, Moon Lake Project, Utah." *Reclamation Era* 27 (March 1937), 66-7.

1635. "The Emergency Conservation Project." *American Forests* 39 (July 1933), 309.

1636. Engbeck, Joseph H. *By the People, For the People: The Work of the Civilian Conservation Corps in California State Parks, 1933-1941.* Sacramento, CA: California State Parks, 2002.

1637. Fairley, Lincoln. "The Civilian Conservation Corps on Mt. Tamalpais, 1933-1940." *Californians* 1 (July-August 1983), 21-7.

1638. Fiock, L.R. "Civilian Conservation Corps Boys in Campaign to Control Gophers." *Reclamation Era* 28 (July 1938), 134-5.

1639. _____. "Work of the Civilian Conservation Corps at Elephant Butte Reservoir." *Reclamation Era* 28 (January 1938), 16-8.

1640. Fisher, C.C. "Civilian Conservation Corps Aids in Construction of Deschutes Project." *Reclamation Era* 29 (May 1939), 101-3.

1641. Frankland, James. "New Worlds to Conquer: C.C.C. Roads in Oregon Open Up Entrancing Recreational Area." *Oregon Motorist* (June 1935), 5, 18.

1642. Gavin, Ava Hays. "Civilian Conservation Corps at Alcova Reservoir." *Reclamation Era* 29 (May 1939), 100-1.

1643. Golze, Alfred R. "CCC On Federal Reclamation Projects." *Civil Engineering* 9 (November 1939), 654-6.

1644. _____. "Civilian Conservation Corps Accomplishments on Federal Reclamation Projects, Fiscal Year 1938." *Reclamation Era* 28 (September 1938), 190-2.

1645. _____. "Civilian Conservation Corps Accomplishments on Federal Reclamation Projects, Fiscal Year 1939." *Reclamation Era* 29 (September 1939), 248-51.

1646. _____. "Civilian Conservation Corps Accomplishments on Federal Reclamation Projects, Fiscal Year 1940." *Reclamation Era* 30 (November 1940), 318-20.

1647. _____. "Civilian Conservation Corps Accomplishments on Reclamation Projects." *Reclamation Era* 27 (January 1937), 20-1.

1648. _____. "Civilian Conservation Corps and Emergency Work." *Reclamation Era* 27 (August 1937), 188-9.

1649. _____. "Emergency Conservation Work Activities." *Reclamation Era* 26 (April 1936), 102-3.

A short article that tells of the work activities of CCCers. It also mentions enrollees participation on local radio stations.

1650. _____. "Reclamation and Emergency Conservation Work." *Reclamation Era* 25 (December 1935), 235-6+.

This article gives a good description of the make up of a Bureau of Reclamation CCC camp and summarizes how the CCC enrollees have assisted the Bureau of Reclamation.

1651. _____. "Summer Program of Reclamation C.C.C. Camps." *Reclamation Era* 27 (April 1937), 86-7.

This article lists locations of camps by project name and camp number between April 1, 1937 and September 30, 1937.

1652. Goodwin, T.R. "Removing the 'Death' from Death Valley." *American Forests* 41 (May 1935), 215-8+.

1653. Goodwin, Theodore R. "How the CCC's Built Death Valley National Monument." *Oasis* (1935), 2.

1654. Hafenrichter, A.L. "Lassoing the West's Rampaging Dunes." *Yearbook of Agriculture 1967*, 317-21.

1655. Imrie, G.C. "Construction of Pine View Dam." *Reclamation Era* 27 (August 1937), 180-3.

1656. Jensen, E.S. "Civilian Conservation Corps Constructs Midview Dam, Moon Lake Project, Utah." *Reclamation Era* 38 (July 1938), 136-8.

1657. Johnson, Leon G. "This Work Must Go On." *Journal of Forestry*, 32 (December 1934), 933-7.

Johnson talks about his experiences as a Project Superintendent at Tahoe National Forest in California.

1658. Kay, R.C. "Civilian Conservation Corps Reconstructs Alkali Creek Inclined Drop, Shoshone Project." *Reclamation Era* 31 (January 1941), 11-2.

This article describes CCC participation on the Alkali Creek, which was part of the Shoshone Project. The new inclined water drop was constructed over the old existing one, which was beginning to disintegrate.

1659. Kramer, E.W. "Engineering Activities of CCC Camps in California: Bridge Building, Road Construction, and Control of Floods and Erosion." *Civil Engineering* 4 (July 1934), 335-9.

Kramer tells of the first six months work activities of the CCC in California. Pictures show many of the bridges constructed and the locations of dam projects.

1660. Lumpee, Henry L. "The Civilian Conservation Corps: Irrigator." *Reclamation Era* 26 (November 1936), 268-9.

Lumpee talks about the Vale, Oregon Project. The CCC completed the project with construction of the Willow Creek lateral system.

1661. Mannick, Alfred. "Civilian Conservation Corps Enrollees Reconstructs Portion of Mabton Siphon, Yakima Project." *Reclamation Era* 29 (July 1939), 181-2.

1662. _____. "Civilian Conservation Corps Reconstruction of Pipe Lines, Sunnyside Division, Yakima Project, Washington." *Reclamation Era* 31 (April 1941), 108-9.

1663. Martin, Melvin R. "Civilian Conservation Corps Cleating Operations in Shasta Reservoir." *Reclamation Era* 31 (May 1941), 144-6.

1664. Maxey, William, Jr. "Civilian Conservation Corps Activities on the Milk River Project, Montana." *Reclamation Era* 29 (October 1939), 257-9

1665. Nelsen, Clair E. "Remembering the CCC: Buck Meadows, California, 1933-1934." *Journal of Forest History* 26 (October 1982), 184-91.

1666. Ober, Michael J. "The CCC Experience in Glacier National Park." *Montana, The Magazine of Western History* 26 (Summer 1976), 30-9.

1667. Olsen, F.J. "Civilian Conservation Corps Goes to War on Rodents." *Reclamation Era* 31 (July 1941), 197-8.

1668. Parkinson, Dana. "The CCC Helps Farmers in Utah." *The Utah Farmer* (March 25, 1935), 10.

1669. Pinkerton, John B. "The Youth Development and Conservation Corps in the State of Washington." *Journal of Forestry* 62 (June 1964), 373-5.

1670. Shepard, Earl W. "Civilian Conservation Corps Reconstructs McMillan Dam and Avalon Dam Spillway, Carlsbad Project." *Reclamation Era* 29 (April 1939), 72-3+.

1671. "Supervisor and Five C.C.C. Men Lose Lives in Western Fires." *American Forests* 40 (October 1934), 482.

1672. Swain, Donald C. "The Bureau of Reclamation and the New Deal, 1933-1940." *Pacific Northwest Quarterly* 61 (July 1970).

1673. _____. "The National Park Service and the New Deal, 1933-1940." *Pacific Historical Review* 41 (August 1972), 312-32.

1674. Tanasoca, Donald. "Six Months in Garden Valley." Ed. by Elmo Richardson. *Idaho Yesterday* 11 (Summer 1967), 16-24.

1675. Throop, Elizabeth Gail. "A Characteristic Expression: A Thematic Evaluation of Forest Service Depression-Era Administrative Buildings in the Pacific Northwest." *Contract Abstracts and CRM Archaeology* 3 (1983), 123-9.

1676. _____. "Utterly Visionary and Chimerical: A Federal Response to the Depression: An Examination of Civilian Conservation Corps Construction on National Forest System Lands in the Pacific Northwest." Master's thesis, Portland State University, 1979.

1677. Weber, R.C.E. "Emergency Conservation Work, Yuma Project, Arizona-California." *Reclamation Era* 26 (May 1936), 124-5.

1678. Wilkinson, S. Kristina and Michael R. Hardwick. "La Purisima, A Living Museum of Spanish Colonial History." *Noticias* 9:2 (1973), 1-9.

1679. Williamson, Thomas. "Civilian Conservation Corps Construction of Parapet and Curb Walls, Rye Patch Dam, Humboldt Project, Nevada." *Reclamation Era* 29 (January 1939), 15-6.

1680. Woods, Wilma M. "The Role of Federal Dollars and the Economics of Recreation in the Development of Idaho's State Parks, 1908-1965." *Idaho Yesterday* 34:3 (1991), 15-21.

NATIONAL DEFENSE

1681. Abbot, H.D. "National Defense and the CCC." *Soil Conservation* 6 (November 1940), 117-9.

1682. Brown, G.S. "CCC Camps as Red Cross Evacuee Centers." *American City* 57 (January 1942), 37.

This article discusses and arrangement between the Red Cross and the CCC, where CCC personnel, equipment and facilities would be used by the American Red Cross if necessary in the early days of WW II.

1683. Butler, Ovid. "Should the Civilian Conservation Corps be Liquidated as Unessential to National Defense?" *American Forests* 48 (February 1942), 53.

1684. "CCC and Defense." *American Forests* 47 (July 1941), 333.

1685. "CCC's Contribution to the War Effort and to American Education." *School and Society* 55 (June 20, 1942), 688.

1686. "CCC Needed for Service in Defense." *American Forests* 47 (September 1941), 405.

1687. "CCC Repair Shops Transferred to Army." *American Forests* 48 (May 1942), 231.

1688. "CCC to Intensify Defense Training." *American Forests* 47 (July 1941), 345.

1689. Culley, John Joel. "A Troublesome Presence: World War II Internment of German Sailors in New Mexico." *Prologue* 28 (Winter 1996), 278-95.

Culley tells of German merchant sailors who were taken off a sinking ship and brought top the United States and interred in a former CCC camp. This was before the U.S. and Germany were at war. They became the first civilians interred during WW II.

1690. "Evacuation: CCC Camps for Emergency Use." *Survey* 78 (January 1942), 18.

1691. "Fortune Survey: Would You Favor Giving Military Training to the CCC Boys?" *Fortune* 20 (December 1939), 116.

1692. Golze, Alfred R. "Operation and Maintenance of Civilian Conservation Corps Equipment on Reclamation Projects--Its Relation to National Defense." *Reclamation Era* 31 (April 1941), 118-20.

1693. Guthrie, John. "Forestry in National Defense." *Journal of Forestry* 40 (June 1941), 477-80.

1694. Hagood, Johnson. "Soldiers of the Shield: The Civilian Conservation Corps Stands Square With National Defense." *American Forests* 40 (March 1934), 103-5.

1695. McEntee, James J. "CCC Aids National Defense." *Black Hills Engineer* 26 (December 1940), 173-80.

1696. _____. "CCC and National Defense." *American Forests* 46 (July 1940), 309-10.

1697. Oxley, Howard W. "CCC and the Defense of the Nation." *Education* 61 (October 1940), 74-7.

1698. _____. "Training for National Defense." *School Life* 26 (July 1941), 315-6.

1699. Shoemaker, Blair. "The CCC and Preparedness." *American Forests* 45 (July 1939), 338.

1700. "Specific Training for Non-Combatant Defense Service." *Survey* 76 (December 1940), 362.

SPECIAL EDITION JOURNALS

1701. *American Civic Annual,* Published by the American Civic Association, Inc., Union Trust Building, Washington, DC. This annual touts itself as "A record of recent civic advance with a list of who's who in civic achievement among the members of the American Civic Association." Issues between 1934-1941 contain information regarding activities of the CCC. The table of contents of each volume is presented by national, regional, state and city issues.

1702. *Black Hills Engineer*, December 1937 (vol. 24, no. 1), Published by the South Dakota School of Mines, Rapid City, SD

1703. Fechner, Robert. "The Civilian Conservation Corps." p. 3.

1704. Krueger, Theodore. "The CCC in the Black Hills and Harney National Forest." p. 14.

1705. Freeland, E.D. "Development of National Park Areas in the Black Hills by the CCC." p. 26.

1706. Powell, Ben F. "Construction of Forest Service Dams by CCC Forces." p. 33.

1707. Averill, C.C. The Civilian Conservation Corps as a Fire Suppression Organization." p. 38.

1708. Roeser, J., Jr. "The Role of Timber Stand Improvements in the Black Hills." p. 48.

1709. Morris, Ralph N. "The Project Superintendent's Duties and Responsibilities in CCC Work." p. 55.

1710. Kent, Leon F. "The Army and the CCC." p. 63.

1711. Krier, Adrian W. "The Educational Program and the Educational Adviser in the Black Hills." p. 68.

1712. Martin, Harold. "Recreational and Conservational Facilities." p. 78.

1713. Anderson, P.C. "Work Accomplished by Camp Narrows in Custer State Park During the Past Three Years." p. 80.
1714. Editorial. p. 82.

1715. *CRM (Cultural Resource Management)*,Published by the National Park Service, Cultural Resources Division, Washington, DC. Volume 19, No. 9 (1996) Supplement is titled National Register of Historic Places Multiple Property Submissions. Examples of listings include CCC Historic Properties in Alaska MPS (Multiple Properties Submission); ECW Architecture in Missouri State Parks 1933-1942 TR (Thematic Resource); and, New Deal Construction in Kentucky MRA (Multiple Resource Area).

1716. *Education*, October 1940 (vol. 61, no. 2), Published by The Palmer Company, Boston, "Devoted to The Science, Art, Philosophy, and Literature of Education."
1717. Fox, John A. "The Challenge to CCC." pp. 65-8.
1718. Green, William. "Labor and the Youth Situation." pp. 68-70.
1719. Judd, Charles H. "The Federal Government and Youth." pp. 70-4.
1720. Oxley, Howard W. "The CCC and the Defense of the Nation." pp. 74-7.
1721. Bryan, John E. "Youth Learn Manual Skills." pp. 78-82.
1722. Lukens, Fred E. "The CCC and Schools." pp. 82-7.
1723. Sloper, Kirkland. "Guidance Possibilities of the CCC." pp. 88-90.
1724. Lanigan, James S. "Education in the CCC: Weapon or Feint." pp. 91-5.
1725. Wiedman, Dueffort E. "Citizenship Through Training in the Civilian Conservation Corps." pp. 95-6.
1726. Baurer, Herbert T. "Some Effects of Short-Term Trade Experience Classes Upon the Stenquist Mechanical Aptitude Test." pp. 97-100.
1727. Scully, James H. and Harold J. Mahoney. "A Study of Reading Ability in CCC Camps." pp. 101-7.

1728. *Forestry News Digest*, Published by the American Tree Association, Washington, DC, later by the Charles Lathrop Pack Forestry Foundation, Washington, DC.
1729. January 1934--Special CCC Work Report Edition
1730. June 1934--Special Civilian Conservation Corps Edition
1731. November 1934--CCC State Reports Edition

1732. *Foxfire*, Winter 1982 (vol. 16, no. 4), Published by the students of Rabun County High School, Clayton, GA.
1733. Cilbrith, Tallie. "The CCC: The Road to Recovery." p. 228.
1734. Shope, Scott. "James Cartwright CCC Forester." p. 236.
1735. "An Outline for a CCC-Sponsored Course." p. 258.
1736. "The CCC is the Best Thing that ever Happened During the Depression." p. 265.
1737. "The Second Annual Gathering of the Veterans of Camp F-1." p. 278.

1738. *Foxfire*, Spring 1983 (vol. 17, no. 1)
1739. "The Depression." p. 4.
1740. "Enrollment - Camps." p. 8.
1741. "The Specific Jobs." p. 26.
1742. "Food - The Mess Hall." p. 40.
1743. "Recreation - Free Time." p.48.
1744. "Education." p. 54.
1745. "Punishment." p. 56.
1746. "Accidents." p. 58.
1747. "The Final Analysis." p. 61.

1748. *Indians at Work*, 1933 thru 1942, Published by the Bureau of Indian Affairs, U.S. Department of the Interior, Washington, DC. Almost every issue contains information pertaining to the Native Americans' work in the CCC. Tells of specific tribal projects and the progress made.

1749. *Journal of Forestry*, March 1936 (vol. 34, no. 3), Published by the Society of American Foresters, Washington, DC. The March issue contains papers presented at the January 29, 1936, session at the 35th Annual Meeting of the Society of American Foresters, in Atlanta, GA.
1750. Morrell, Fred. "Emergency Conservation Work Policies." p. 284-92.
1751. Alexander, Thomas W. "Comments on the C.C.C." p. 292-7.
1752. Jeffers, D.S. "Further Comments on the C.C.C." p. 298-302.
1753. "Statement Supplementary to the 1935 Report of the C.C.C. Committee of the Society." p. 302-7.
1754. "Report on the Civilian Conservation Corps." p. 307-19.

1755. *Phi Delta Kappan*, May 1937 (vol. 19, no. 9), Published by the Ovid Bell Press, Inc., for *Phi Delta Kappa,* a professional fraternity in education, Chicago, IL.
1756. Editorial Comment.

1757. Studebaker, John W. "Possibilities for Education Through the CCC." p. 297.
1758. Judd, Charles H. "Educational Activities in the CCC." p. 299.
1759. Dearborn, Ned H. "Educational Opportunity for Enrollee." p. 300.
1760. Hill, Frank Ernest. "Laboratory for Adult Education." p. 302.
1761. Cook, Paul M. "Permanency of the CCC." p. 304.
1762. Fechner, Robert. "The Educational Contribution of the Civilian Conservation Corps." p. 305.
1763. [Conley, E.T.] "The Adjutant General Emphasizes Importance of Educational Program." p. 308.
1764. "It May Interest You to Know That..." p. 310.
1765. Oxley, Howard W. "Growth and Accomplishments of CCC Education." p. 313.
1766. Holland, Kenneth. "Education in CCC and European Camps." p. 317.
1767. "That Poignant 41.09 Per Cent." p. 322.
1768. Persons, W. Frank. "Human Resources and the Civilian Conservation Corps." p. 323.
1769. Leavitt, Scott. The Social and Economic Implications of Conservation." p.326.
1770. Sellers, Sandford, Jr. "The Corps Area Office and Its Relation to the Camp Educational Program." p. 328.
1771. Williams, Paul E. "Community and Camp Work Together for Guidance and Adjustment." p. 334.
1772. Beam, Russell A. "Counseling for Adjustment and Rehabilitation." p. 337.
1773. Morrell, Fred. "Job Training in the C.C.C." p. 342.
1774. Nystrom, Joel E. "Informal Educational Activities." p.346.
1775. Rogers, L.W. "Training in Vocational Fundamentals." p.351.
1776. Blanchard, C.H. "I Talk with My CCC Boys." p. 354.
1777. Boutwell, William Dow. "The Use of Radio in CCC Camps." p. 356.
1778. Wiedman, D.E. "Co-operative Relations Between the CCC and Educational Institutions." p. 358.
1779. Standley, L.L. "Eye Tests in the Promotion of Safety." p. 359.
1780. Burke, Joseph W. "Correspondence Work in the Camps of the Ninth Corps Area." p. 361.
1781. Reed, A.A. "Correspondence Work Offered by the University of Nebraska for Students in the CCC." p. 363.
1782. Stevenson, Fred G. "Michigan Supervised Correspondence Study." p. 365.

1783. Oxley, Howard W. "Research Projects in CCC Education." p. 368.
1784. "Annotated Bibliography: The Civilian Conservation Corps." p. 350.
1785. "Recent Legislation Concerning the CCC." p. 375.
1786. "Keeping Abreast of the Times." p. 379.

1787. *Soil Conservation*, March 1938 (vol. 3, no. 9). Published by the U.S. Department of Agriculture, Soil Conservation Service, Washington, DC.
1788. Bennett, H.H. "The Opportunity Broadens." p. 235.
1789. Fechner, Robert. "Youth and the Land." p. 236.
1790. Abbot, H.D. "CCC Operations." pp. 237-8.
1791. Tyner, George P. "The Enrollee and Soil Conservation." p. 239.
1792. Persons, W. Frank. "A Selecting Agency Views C.C.C. Work Opportunities." pp. 240-1.
1793. Giesler, Cecil O. "How I Plan to Use Experience Acquired in an S.C.S./C.C.C. Camp." pp. 242+.
1794. Seaman, James A. "Enrollee Aid in Northeastern Orchards." p. 243.
1795. Howard, Ivy M. "Pioneers of Today." pp. 244-6.

1796. *Transactions of the North American Wildlife Conference.* Published by the American Wildlife Institute, Washington, DC. Each annual edition between 1936 and 1951 contains materials related to the Civilian Conservation Corps.

SELECTED GOVERNMENT DOCUMENTS

Note: It should be assumed that documents in this section were printed in Washington, DC, by the Government Printing Office (GPO), unless otherwise noted.

1797. United States. Civilian Conservation Corps. *The CCC and Its Contribution to A Nation-Wide State Park Recreational Program.* Washington: GPO, 1937.

1798. _____. *The Civilian Conservation Corps and Colored Youth.* Washington: CCC, 1940.

1799. _____. *The Civilian Conservation Corps and Public Recreation.* Washington: GPO, 1941.

1800. _____. *Civilian Conservation Corps Bibliography: A List of References on the United States Civilian Conservation Corps.* Washington: Office of the Director, CCC, 1939.

1801. _____. *The Civilian Conservation Corps, "Builder of Men."* Washington: Happy Days Publishing Co., Inc., 1940.

1802. _____. *Civilian Conservation Corps. Contributing to the Defense of the Nation.* Washington: GPO, 1941.

1803. _____. *The Civilian Conservation Corps: What it is and What it Does.* Washington: Office of the Director, CCC, 1939.

1804. _____. *Emergency Conservation Work (Civilian Conservation Corps): Fourth Anniversary Report to the President.* Washington: CCC, 1937.

1805. _____. *Hands to Save the Soil.* Washington: GPO, [1939].

1806. _____. *Memories of Civilian Conservation Corps, 1941 (Yearbook).*

1807. _____. *Official Annual, 1937, Civilian Conservation Corps, First CCC District, First Corps Area.* Baton Rouge, LA: Direct Advertising Co., 1937.

1808. _____. *Official Annual, Civilian Conservation Corps: Nebraska-South Dakota District, CCC Seventh Corps Area.* Baton Rouge, LA: Direct Advertising Co., 1937.

1809. _____. *Pictorial Review, Civilian Conservation Corps, District I, Company 2444, FLA. SP-5.* Washington: GPO, 1939.

1810. _____. *Proceedings, C.C.C. Foreman Training Conference* (Washington, May 13-25, 1940). Washington(?): Federal Security Agency, CCC, 1940.

1811. _____. *Project Training Series.*

1812. No. 1. *The "Why" of Block-and-Tackle.*
1813. No. 2. *Outline of Elementary Courses in Forestry.*
1814. No. 3. *Carpentry Joints and Splices.*
1815. No. 4. *Concrete.*
1816. No. 5. *Brick and Stone Work.*
1817. No. 6. *Landscape Conservation.*
1818. No. 7. *Construction of Trails.*
1819. No. 8. *Lumber.*
1820. No. 9. *Signs and Markers.*
1821. No. 10. *Construction of Relief Models.*
1822. No. 11. *Carpentry--Small Frame Buildings.*
1823. No. 12. *Carpentry--Four Room Frame House.*
1824. No. 13. *Common Range Plants.*
1825. No. 14. *Lawns.*
1826. No. 15. *Job Training is a Business Proposition.*
1827. No. 16. *Job Training is in Work.*
1828. No. 17. *Responsibility for Job Training.*
1829. No. 17-A. *Appendix to No. 17.*
1830. No. 18. *Job Training Methods.*
1831. No. 19. *A Statement of Principles.*
1832. No. 20. *Proceedings. C.C.C. Foreman Training Conference-* Washington, May 13-25, 1940.
1833. No. 21. *Concrete Contracting.*

1834. _____. *Standards of Eligibility and Selection for Junior Enrollees.* Washington: GPO, 1939.

1835. _____. *Studies in CCC Education: Papers Prepared for Conference of Education Advisors, June 3-6, 1940.* Governor's Island, NY: CCC, Office of Education, Second Corps Area, 1940.

1836. United States. Congress. House. Committee on Labor. *To Make the Civilian Conservation Corps A Permanent Agency.* 76th Congress, First Session, H.R. 2990. Washington: GPO, 1937.

1837. United States. Department of Agriculture. Bureau of Biological Survey. *The CCC and Wildlife.* Washington: GPO, 1939.

1838. _____. Forest Service. *Camp Stoves and Fireplaces* by A.D. Taylor. Washington: Emergency Conservation Work, 1937.

1839. _____. *The Civilian Conservation Corps' (CCC) Contribution to the Forests in the Pacific Northwest Region.* (Portland): USDA Forest Service, Pacific Northwest Region, 1989(?).

1840. _____. *The Civilian Conservation Corps: Coronado National Forest, 1933-1942.* Washington: GPO, 1991.

1841. _____. *The Forest Service and the Civilian Conservation Corps: 1933-1942.* Washington: GPO, 1986.

1842. _____. Forestry Publications Series.

1843. No. 1. *Measures for Stand Improvement in Southern Appalachian Forests*--Appalachian Forest Experiment Station, June 1933.
1844. No. 2. *Eastern Forest Tree Disease in Relation To Stand Improvement*--George W. Hepting, Bureau of Plant Industry, Revised may 1940.
1845. No. 3. *Stand-Improvement Measures for Southern Forests*--Southern Forest Experiment Station, November 1933.
1846. No. 4. *Contour-Trenches Control Floods and Erosion on Range Lands*--Reed V. Bailey and A.R. Croft, Intermountain Forest and Range Experiment Station, May 1937.
1847. No. 5. *Collecting and Handling Seeds by Wild Plants*--N.T. Mirov and Charles J. Kraebel, California Forest and Range Experiment Station, June 1939.
1848. No. 6. *Timber Stand-Improvement in the Southwest*--G.A. Pearson, Southwest Forest and Range Experiment Station, 1940.
1849. No. 7. *Reforestation in the Southwest by CCC Camps*--G.A. Pearson, Southwest Forest and Range Experiment Station, 1940.
1850. No. 8. *Revegetating Semi-Desert Range Lands in the Southwest*--John T. Cassady and George T. Glendening, 1940.

1851. No. 9. *Diseases of Forest Tree Nursery Stock*--W.C. Davis, Bureau of Plant Industry, 1942.

1852. _____. *Improvement Handbook*, by the Division of Engineering. Washington: GPO, 1937.

1853. _____. *Acceptable Plans for Forest Service Administrative Buildings*, by the Division of Engineering. Washington: GPO, 1938.

1854. _____. *Forest Administrative Site Plans.* Region 6. 1935-1940. (Washington: GPO).

1855. _____. *The Rise of Multi-Use Management in the Intermountain West: A History of Region 4 of the Forest Service*, by Thomas G. Alexander. Salt Lake City, UT: Forest Service, 1987.

1856. United States. Department of the Interior. *Annual Report, Secretary of the Interior.* Washington: GPO, 1933-1942.

1857. _____. *Reports on Cooperation Between Tennessee Valley Authority and the Bureau of Reclamation, Geological Survey and National Park Service of the Department of the Interior.* Washington: GPO, 1938.

1858. _____. Bureau of Reclamation. *Civilian Conservation Corps Handbook.* Washington: Department of the Interior, 1933-42.

This handbook is a compilation of official correspondence and instructions for the CCC camps under the direction of the Bureau of Reclamation. Forms and instructions were added as amended. This copy is housed in the Department of the Interior Library in Washington.

1859. _____. *Reclamation Project Data.* Washington: GPO, 1961.

1860. _____. Emergency Conservation Work. *Landscape Conservation.* Amherst, MA: Massachusetts State College, 1935 (Project Training Series No. 6).

1861. _____. *Interim Report of [Director of] Emergency Conservation Work, Apr. 5-Dec. 31, 1933.* Washington: January 25, 1934.

1862. _____. Indian Division. *Civilian Conservation Corps–Indian Division Handbook.* Washington: Department of the Interior, [1941].

This handbook contains all reporting forms and instructions for operating an Indian Division CCC camp. It is an administration manual, which also contains photographs and drawings for construction projects. It is located at the Department of the Interior Library in Washington.

1863. _____. National Park Service. *Administration Manual for Recreational Demonstration Areas.* Washington: Department of the Interior, 1941.

1864. _____. *The CCC and its Contribution to a Nation-Wide State Park Recreational Program.* Washington: GPO, 1937.

1865. _____. *Emergency Conservation Work Handbook.* (No. 182). Washington: Department of the Interior, n.d.

1866. _____. *Master Plans: A Manual of Standard Practice for Use in the National Park Service.* Washington: National Park Service, 1939.

1867. _____. *Parkways: A Manual of the Revised Requirements, Instructions, and Information Relating to National Parkways for Use in the National Park Service.* Washington: National Park Service, 1938.

1868. _____. *Portfolio of Comfort Stations and Privies.* Washington: National Park Service, 1934.

1869. _____. *Portfolio of Housekeeping Cabins.* Washington: National Park Service, Branch of Plans and Designs, n.d.

1870. _____. *Portfolio of Park Structures and Facilities.* [Washington: GPO], 1934.

1871. _____. *A Study of the Park and Recreation Problem of the United States.* Washington: GPO, 1941.

1872. _____. State Park Emergency Conservation Work. *Soil Conservation Work: Causes, Influencing Factors, and (Suggested) Methods of Control Applicable to Park Areas in the Seventh Region ECW.* n.p., 1935.

1873. _____. Technical Circulars. No. 1. *The Construction of Relief Models.* [Washington: GPO], 1936.

1874. _____. *Tree Preservation Bulletins.* Branch of Forestry.

1875. No. 1 & 2. *Purpose and Policy*, December 1935, Revised March 1938.
1876. No. 3. *Tree Bracing*, December 1935, Revised April 1938.
1877. No. 4. *Shade Tree Pruning*, February 1936, Revised January 1938.

1878. No. 5. *Lightning Protection for Trees*, April 1936, Revised April 1938.

1879. No. 6. *General Spraying Practices*, May 1936, Revised April 1938.

1880. No. 7. *Rope, Knots and Climbing*, August 1936, Revised April 1938.

1881. No. 8. *Safety for Tree Workers*, October 1937.

1882. No. 9. *Transplanting Trees and Other Woody Plants*, January 1940.

1883. United States. Department of Labor. Emergency Conservation Work Bulletins.

1884. No. 1. *Chance to Work in Forests, Questions and Answers for Information of Men Offered Opportunity to Apply for National Emergency Conservation Work* (with revisions). Washington: 1933.

1885. No. 2. *National Emergency Conservation Work, What it is and How it Operates*. Washington: 1933

1886. No. 3. *Handbook for Agencies Selecting Men for Emergency Conservation Work* (with revisions). Washington: 1933.

1887. _____. *CCC: Standards of Eligibility and Selection*. Washington: GPO, 1938.

1888. _____. *Selecting 1,800,000 Young Men for the CCC*, Serial No. R 745. Washington: GPO, 1938.

1889. _____. *The Selection of Men for the Civilian Conservation Corps. A Narrative Record Covering the Work of the United States Department of Labor During Two Years-April 5, 1933, to April 5, 1935*. [Washington]: 1935.

1890. _____. *What is Emergency Conservation Work?* Washington: GPO, 1935.

1891. United States. Emergency Conservation Work. *Architecture in Missouri State Parks. Thematic Resources*. Washington: GPO, 1984.

1892. _____. *Histories of Southeastern State Park Systems*. Washington: GPO, 1977.

1893. _____. Safety Division Bulletins ECW 1-ECW 23.

1894. ECW 1. *Transportation*. August 30, 1934. 2 pp.

1895. ECW 2. *Guard-Rails on Dump Trucks*. August 30, 1934. 1 p.

1896. ECW 3. *Swimming*. June 22, 1934. 1 p.
Supplement: August 15, 1934. 1 p.

1897. ECW 4. *Snake-Bite.* June 22, 1934. 2 pp.
1898. ECW 5. *Goggles.* June 28, 1934. 1 p.
Supplement: (*Goggle Specifications*) August 15, 1934. 2 pp. Also Carries Heading--Adjutant General's Office, War Department.
1899. ECW 6. *Poison Ivy and Poison Sumac.* July 7, 1934. 2 pp.
1900. ECW 7. *Sunstroke and Heat Exhaustion.* July 9, 1934. 1 p.
1901. ECW 8. *Scaffolds.* July 13, 1934. 1 p.
1902. ECW 9. *First Aid Classes.* July 24, 1934. 2 pp. Supplement: December 11, 1934. 1 p.
1903. ECW 10. *Excavation and Trench Construction.* August 8, 1934. 2 leaves.
1904. ECW 11. *Electrical Hazards.* August 15, 1934. 2 pp.
1905. ECW 12. *Rules for Handling Explosives.* August 21 (or 30), 1934. 14 pp. (ECW 12, August 30, 1934, Rescinded by ECW 19, November 30, 1934).
1906. ECW 13. *Use of Water Crafts by Enrolled Men of Emergency Conservation Work.* August 23, 1934. 1 p. (Rescinded by CW 16, September 17, 1934).
1907. ECW 14. *Chiggers and Ticks.* August 24, 1934. 2 p.
1908. ECW 15. *Fire Prevention "Don'ts."* September 12, 1934. 1 p.
1909. ECW 16. *Use of Water Craft by Enrolled Men of Emergency Conservation Work.* September 17, 1934 1 p.
(Remands ECW 13).
1910. ECW 17. *Urgent and Important. Winter Months.* October 12, 1934. 1 p.
1911. ECW 18. *Suggested Safety Policy and Procedure.* November 13, 1934. 3 leaves.
1912. ECW 19. *Guard-Rails and Portable Seats on Trucks.* November 30, 1934. 1 p. (Reminds ECW 12).
1913. ECW 20. *Fire Protection, Gasoline and Oil Hazards.* December 17, 1934. 1 p.
1914. ECW 21. *Fire Prevention and Fire Fighting.* February 5, 1935. 3 leaves.
1915. ECW 22. *Handling and Care of Tools.* February 20, 1935. 2 leaves.
1916. ECW 23. *Life Saving Schools.* May 20, 1935.

1917. _____. *Safety Division Regulations, Civilian Conservation Corps. Part I. Part II. Part III. Mandatory Rules and Regulations. Specifications. Suggested Safety Material.* Washington: GPO, 1935.

1918. _____. *Two Years of Emergency Conservation Work (Washington: Civilian Conservation Corps) April 5, 1933-March 31, 1935.* Washington: Report Prepared by Robert Fechner, 1935.

1919. United States. Federal Security Agency. Civilian Conservation Corps. *The CCC at Work: A Story of 2,500,000 Young Men.* Washington, GPO, 1941.

The CCC at Work is heavily illustrated with one page explanations of the daily activities of the CCC. Included are work projects, educational and recreational activities.

1920. _____. *The CCC and National Defense.* Washington: GPO, 1941.

1921. _____. *Civilian Conservation Corps Bibliography: A List of Magazine References and Books on the United States Civilian Conservation Corps.* Washington: Office of the Director, Civilian Conservation Corps, 1943.

1922. _____. *Civilian Conservation Corps: A Brief Summary of Certain Phases of the CCC Program--Missouri, April 1933-June 30, 1942.* n.p. 1942.

1923. _____. *Education in the Civilian Conservation Corps.* Report Prepared by Howard Oxley. Washington: [GPO], 1936.

1924. _____. Office of Education. Instructional Series.

1925. No. 1. *Photography*
1926. No. 2. *Blacksmithing*
Unit 1. Hand Forging
1927. No. 3. *Arc Welding*
Unit 1. Basic Processing
1928. No. 4. *Bricklaying*
Unit 1. Elementary Bricklaying
1929. No. 5. *Masonry*
Unit 1. Elementary Stone Masonry
1930. No. 6. *Surveying*
Unit 1. Care and Use of Surveying Instruments
1931. No. 7. *General and Household Mechanics*
Unit 1. Household Trouble Shooting
1932. No. 8. n/a
1933. No. 9. *Acetylene Welding*
Unit 1. Basic Processes
1934. No. 10. *Drafting*
Unit 1. Mechanical Principles
1935. No. 11. *Drafting*
Unit 2. Elementary Map Drafting and Reading
1936. No. 12. *Drafting*
Unit 3. Architectural Drawing and Blueprint Reading

1937. No. 13. *Retail Sales*
Unit 1. The Sales Clerk
1938. No. 14. *Retail Sales*
Unit 2. Basic Store Selling
1939. No. 15. *Retail Sales*
Unit 3. Outside Sales
1940. No. 16. *General Agriculture*
Unit 1. Orientation in Agriculture
1941. No. 17. *Poultry Raising*
Unit 1. Starting a New Flock
1942. No. 18. *Poultry Raising*
Unit 2. The Established Flock
1943. No. 19. *Farm Mechanics*
Unit 1. Tool Fitting and General Repairing
1944. No. 20. *Farm Mechanics*
Unit 2. Pore and Construction Work
1945. No, 21. n/a
1946. No. 22. n/a
1947. No. 23. n/a
1948. No. 24. n/a
1949. No. 25. n/a
1950. No. 26. *Service Station Management*
Unit 1. Driveway Service Procedures
1951. No. 27. *Service Station Management*
Unit 2. Lubrication, Sales and Service

1952. _____. Visual Aids Handbook

Unit 1. *The Use of Films, Filmstrips and Slides*

1953. _____. Vocational Training Series.

1954. No. 1. *Agriculture*
1955. No. 2. *Automobile Repairing*
1956. No. 3. *Automotive Electricity*
1957. No. 4. *Carpentry*
1958. No. 5. *Concrete Construction*
1959. No. 6. *Cooking*
1960. No. 7. *Conservation of Natural Resources*
1961. No. 8. *Forestry*
1962. No. 9. *House Wiring*
1963. No. 10. *Elementary Masonry & Bricklaying*
1964. No. 11. *Mechanical Drawing*
1965. No. 12. *Photography*
1966. No. 13. *Radio Servicing*

1967. No. 14. *Soil Conservation*
1968. No. 15. *Plane Surveying*

1969. _____. *Work Experience that Counts.* Washington: [GPO], 1941.

1970. United States. Library of Congress. Directorate of the Legislative Reference Service. Civilian Conservation Corps. 77th Cong.,2nd sess., Senate Document No. 216, 1942.

1971. _____. Division of Bibliography. *A List of References on the United States Civilian Conservation Corps.* Washington: Library of Congress, 1936. And Supplements: June 30, 1937; and, May 6, 1939.

1972. _____. *A Selected List of References on the New Deal.* Compiled by Florence S. Hellman, Chief Bibliographer, June 14, 1940.

1973. United States. National Archives and Records Service. *Cartographic Records of the Bureau of Indian Affairs.* (Special List No. 13) Laura E. Kelsay, comp. Washington: National Archives and Records Service, 1981.

1974. _____. *Cartographic Records of the Forest Service.* (Preliminary Inventory No. 167) Charlotte M. Ashby, comp. Washington: National Archives and Records Service, 1967.

1975. _____. *Cartographic Records of the Office of the Secretary of the Interior.* (Preliminary Inventory No. 81) Laura E. Kelsay, comp. Washington: National Archives and Records Service, 1955.

1976. United States. National Civilian Community Corps. *Great Starts: The 30's, Civilian Conservation Corps--the 60's, Peace Corps--the 90's, National Civilian Community Corps: Your Turn.* Washington, National Civilian Community Corps, 1994.

1977. United States. National Resources Committee. Water Resources Committee. *Low Dams: A Manual of Design for Small Water Storage Projects.* Washington: GPO, 1939.

1978. United States. President. Franklin D. Roosevelt, 1933-1945. *Making Civilian Conservation Corps A Permanent Agency.* 75th Congress, First Session, Doc. 196. Washington: GPO, 1937.

1979. United States. Tennessee Valley Authority. *Civilian Conservation Corps Camps in Tennessee: 1933-1942.* Knoxville, TN: Tennessee Valley Authority Technical Library, 1984.

James (Jim) Jones compiled this listing for the Technical Library, TVA, from other research he conducted on the CCC in Tennessee. He lists each camp, location, and designation, for each enrollment period.

1980. United States. War Department. *Civilian Conservation Corps Regulations*. Washington: GPO, 1937.

1981. Virginia. State Department of Public Welfare. Bureau of CCC Selection. [*Selection of CCC Enrollees*]. Richmond, VA: VA State Department of Public Welfare, 1940.

This small pamphlet provides personnel of the Virginia Selection Bureau with an overall view of the CCC program. It provides information for questions that might be asked by prospective enrollees.

Presidential Executive Orders

1982. No. 6101. Relief of Unemployment Through Performance of Useful Public Work (Appointing Robert Fechner as Director of Emergency Conservation Work, Providing for Advisory Council to Director of Emergency Conservation Work, etc.). April 5, 1933. 1 p.

1983. No. 6106 A. Relief of Unemployment Through Performance of Useful Public Work (Directing Chief of Finance of Army to Receive, Expend, and Account for All Funds Expended under Direction of Director of Civilian [Emergency] Conservation Work). April 10, 1933. 1 p.

1984. No. 6109. Relief of Unemployment Through Performance of Useful Public Work (Fixing Cash Allowance for Members of Conservation Corps at Rate of $30 per Month). April 12, 1933. 1 p.

1985. No. 6126. Administration of Emergency Conservation Work [Concerning Completion of Establishment of Office of Director of Emergency Conservation Work in District of Columbia]. May 8, 1933. 2 pp.

1986. No. 6127. Reimbursement for Supplies Furnished Emergency Conservation Work. May 8, 1933. 1 p.

1987. No. 6129. Administration of Emergency Conservation Work (Concerning enrollment of 25,000 Veterans of World War). May 11, 1933. 1 p.

1988. No. 6131. Relief of Unemployment Through Performance of Useful Public Work (Indian Reservations). May 12, 1933. 1 p.

1989. No. 6131 A. Relief of Unemployment Through Performance of Useful Public Work (Designating Naval Surgeons to be Attached to War Department for Duty with Civilian Conservation Corps). May 13, 1933. 7 pp.

1990. No. 6135. Purchase of National Forest Lands. May 20, 1933. 2 pp.

1991. No. 6136. Extending Application of Act for Relief of Unemployment Through Performance of Useful Public Work to County Parks and Metropolitan Sections of Municipal Parks. May 20, 1933. 1 p.

1992. No. 6144. Amendment of Executive Order No. 6129 Entitled Administration of Emergency Conservation Work (Concerning Enrollment of 25,000 Veterans of World War so as to Include Veterans of Previous Wars). May 24, 1933. 1 p.

1993. No. 6147. Emergency Conservation Work upon Indian Reservations. May 26, 1933. 1 p.

1994. No. 6148. Relief of Unemployment Through Performance of Useful Public Work (Designating Naval Medical and Dental Officers to be Attached to War Department for Duty with Civilian Conservation Corps). May 31, 1933. 4 pp.

1995. No. 6160. Administration of Emergency Conservation Work (Consolidating and Amending Executive Order 6129, May 11, 1933, Concerning Enrollment of 25,000 Veterans of World War, Executive Order No. 6135, May 20, 1933, Purchase of National Forest Lands, and Executive Order No. 6147, May 26, 1933, Emergency Conservation Work Upon Indian Reservations). June 7, 1933. 3 pp.

1996. No. 6169. Relief of Unemployment Through Performance of Useful Public Works (Attaching Such Warrant Officers of Coast Guard not Exceeding 170, as Secretary of Treasury may Determine, to War Department for Duty in the New Public Works Projects of Civilian Conservation Corps). June 15, 1933. 1 p.

1997. No. 6181. Purchase of Lands for National Forests. June 24, 1933. 1 p. (Amends Section 5, Executive Order 6160, Relating to Administration of Emergency Conservation Work).

1998. No. 6200. Administration of Emergency Conservation Work (Concerning Authority and Power of Director of Emergency Conservation Work and Welfare of Enrolled Men). July 11, 1933. 1 p.

1999. No. 6208. Purchase of Forest Lands for Emergency Conservation Work. July 21, 1933. 1 p. (Rescinds Section 5 of Executive Order No. 6160, June 7, 1933, and Executive Order No. 6181, June 24, 1933, and Substitutes the Present Executive Order thereafter).

2000. No. 6223. Administration of Emergency Conservation Work (Authorizing Transfer of Funds to Credit the Department of Agriculture for Completing Acquisition of Land for National Arboretum). July 27, 1933. 1 p.

2001. No. 6237. Purchase of National Park Lands [Authorizing Transfer of Funds to Credit of War Department for Purchase of Lands Necessary for Completion of Great Smokey Mountain National Park]. July 28, 1933. 2 pp.

2002. No. 6265. Purchase of Forest Lands for Emergency Conservation Work. September 5, 1933. 1 p.

2003. No. 6331. Withdrawal of Public Lands for Conservation Camp, California. October 11, 1933. 1 p.

2004. No. 6344. Authorization and Allocation of Funds for Purchase of Lands for Forest Research in Connection with National Arboretum). October 20, 1933. 1 p. Supersedes Executive Order No. 6223, July 27, 1933.

2005. No. 6541. Withdrawal of Public Lands for Flood Control Purposes. December 28, 1933. 1 p.

2006. No. 6542. Authorizing Purchase of Land for Emergency Conservation Work. December 28, 1933. 2 pp.

2007. No. 6571. Authorizing Relief of Certain Medical and Dental Officers of Navy from Duty with the Civilian Conservation Corps. January 23, 1934. 1 p.

2008. No. 6575. Revocation of Executive Order No. 6237, (July 28, 1933). January 24, 1934. 1 p.

2009. No. 6684. Authorizing Purchase or Rental of Land for Emergency Conservation Work. April 19, 1934. 1 p.

2010. No. 6724. Authorizing Purchase or Rental of Land for Emergency Conservation Work. May 28, 1934. 1 p. (Rescinds Executive Order No. 6684, April 19, 1934).

2011. No. 6747. Allocating Funds to Meet Emergency and Necessity for Relief in Stricken Agricultural Areas. June 23, 1934. 1 p.

2012. No. 6766. Allocation of Funds for Emergency Conservation Work in Restoration, Improvement, and Development of Wildlife Refuges. June 29, 1934. 1 p.

2013. No. 6787. Amendment of Executive Order No. 6208, July 21, 1933, Entitled Purchase of Forest Lands for Emergency Conservation Work. June 30, 1934. 1 p.

2014. No. 6910 A. Purchase of Forest Lands for Emergency Conservation Work. December 1, 1934. 1 p.

2015. No. 6910 B. Allocating Funds from Appropriation to Meet Emergency and Necessary for Relief in Stricken Agricultural Areas. June 23, 1934. 1 p. (Revokes Executive Order No. 6793, July 11. 1934).

2016. No. 7029. Allocation of Funds to Department of Agriculture for Emergency Conservation Work. April 30, 1935. 1 p.

2017. No. 7034. Establish Division of Applications and Information [of National Emergency Council], Advisory Committee on Allotments, Works Progress Administration. May 6, 1935. 1 p.

2018. No. 7046. Prescribing Rules and Regulations Relating to Wages, Hours of Work, and Conditions of Unemployment under Emergency Relief Appropriations Act of 1935. May 20, 1935.

2019. No. 7119. Amending Executive Order No. 7046 (May 20, 1935). July 30, 1935. 1 p.

2020. No. 7129. Authorizing Acquisition of Land on Isle Royale for Emergency Conservation Work. August 6, 1935. 2 pp.

2021. No. 7190. Amendment of Executive Order 7129, August 6, 1935, Authorizing Acquisition of Land on Isle Royale for Emergency Conservation Work. September 25, 1935. 1 p.

2022. No. 7195. Including in Competitive Classified Civil Service Positions of Junior Assistant to Technician in Administration of Civilian Conservation Corps. September 26, 1935. 1 p.

2023. No. 7223. Amendment of Executive Order No. 7195, Including in Competitive Classified Civil Service Positions of Junior Assistant to

Technician in Administration of Civilian Conservation Corps. November 9, 1935. 1 p.

2024. No. 7329. Authorizing Acquisition of Land Near Petersburg, VA, for Emergency Conservation Work. March 30, 1936. 1 p.

2025. No. 7334. Increasing Limitation Contained in Clause (f) of Section 1 of Emergency Relief Appropriations Act of 1935. April 3, 1936. 1 p.

2026. No. 7371. Amendment of Executive Order No. 6190, Prescribing Rules and Regulations Relating to Administration of Emergency Conservation Work. May 18, 1936. 1 p.

2027. No. 7418. Authorizing Acquisition of Land Near Yorktown, VA, for Emergency Conservation Work. July 20, 1936. 1 p.

2028. No. 7466. Authorizing Purchase of Lands for Jefferson National Forest. October 7, 1936. 1 p.

2029. No. 7677 A. Entitled Civilian Conservation Corps. July 26, 1937. 1 p.

2030. No. 7717. Amending Executive Order No. 7677 A, July 26, 1937, Entitled Civilian Conservation Corps. September 29, 1937. 1 p.

2031. No. 8133. Further Amending Executive Order No. 7677 A, July 26, 1937, as Amended, Entitled Civilian Conservation Corps. May 15, 1939. 1 p.

2032. No. 8221. Further Amending Executive Order No. 75677 A, July 26, 1937, Entitled Civilian Conservation Corps. August 21, 1939. 1 p.

2033. No. 8673. Designating Robert Fechner Memorial Forest, Virginia. February 5, 1941. 1 p.

2034. No. 8743. Extending Classified Civil Service. April 23, 1941. 2 pp.

FILMS & MOTION PICTURES

NOTE: A II Is the designation I have assigned to Archives II at College Park, MD. The number following A II is the Control Number assigned to each film at College Park.

2035. Maine, University of (Augusta). *In the Public Interest. The Civilian Conservation Corps in Maine.* Color, Sound, 58 min., VHS, 1988.

2036. Reichert, Bruce and Pat Metzler. *Outdoor Idaho. CCC in Idaho.* Boise, ID: Idaho Public Television, Color, Sound, VHS, 30 min., 1999.

2037. Schultz, M.J. *The CCC: Its Time and Legacy.* Manitowoc, WI: Father and Son Productions, Color with B/W photographs, Sound, VHS, 25 min., 2001.

Schultz provides an overview of the CCC program, and interviews former enrollees of the Allegheny National Forest in north central Pennsylvania.

2038. Time, Inc. *America's Youth.* B/W, Sound, 19 min., 35mm, 1940. vol. 6, no. 9. (A II, #NWDNM(m)-200-MT-6.9)

2039. _____. *CCC.* B/W, Sound & Silent, 66 min., 35mm, 1935. (A II, #NWDNM(m)-200-MT--T-169)

2040. _____. *Dixie, U.S.A.* B/W, Silent, 80 min., 35mm, 1939. (A II, #NWDNM(m)-200-MT-T-691)

2041. _____. *The March of Time.* B/W, Sound, 20 min., 35mm, 1937. vol. 4, no. 1 (September 10, 1937). (A II, #NWDNM(m)-200-MT-4.01)

2042. _____. *The March of Time.* B/W, Sound, 20 min., 35mm, 1938. vol. 4, no. 8 (March 18, 1938). A II, #NWDNM(m)-200-MT-4.08)

2043. _____. *Paramount News.* B/W, Sound, 9 min., 35 mm, 1941. (A II, #NWDNM(m)-200-PN-1.16)

2044. United States. American Council on Education. *Civilian Conservation Corps.* Filmstrip & Script. Washington, DC: 1944.

2045. United States. Army. Signal Corps. *Army Training and Civilian Conservation.* B/W, Silent, 37 min., 35mm, 1941. (A II, #NWDNM(m)-111-M-651)

2046. _____. *Misc. Film No. 58 (Originally designated Historical Film, No. 1125).* B/W, Sound, 28 min., 35mm, 1935. (A II, #NWDNM(m)-111-M-58)

2047. _____. *New Members of Roosevelt Cabinet...* B/W, Sound, 38 min., 35mm, 1936. (A II, #NWDNM(m)-111-H-1546)

2048. _____. *Prelude to War.* B/W, Sound, 53 min., 35mm, 1942. (A II, #NWDNM(m)-111-OF-1)

2049. _____. *Roosevelt "Roughs It" at Forest Camp.* B/W, Sound, 11 min., 35mm, 1937. (A II, #NWDNM(m)-111-M-800A)

2050. United States. Department of Agriculture. Forest Service. Federal Extension Service. *Bridge Building With the CCC.* B/W, Sound, 17 min., 35mm, 1939. (A II,#NWDNM(m)-33.585)

2051. _____. *Camp Roosevelt.* B/W, Silent, 12 min., 35mm, 1936. (A II, #NWDNM(m)-22.458)

2052. _____. *The Civilian Conservation Corps.* B/W, Silent, 48 min., 35mm, 1936. (A II, #33.538)

2053. _____. *The Civilian Conservation Corps at Work--Arizona.* B/W, Silent, 15 min., 35mm, 1934. (A II, #33.474)

2054. _____. *The Civilian Conservation Corps at Work--Colorado, Wyoming, and South Dakota.* B/W, Silent, 16 min., 35mm, 1934. (A II, #33.475)

2055. _____. *The Civilian Conservation Corps at Work--Lake States.* B/W, Silent, 33 min., 35mm, 1936. (A II, #NWDNM(m)-33.487)

2056. _____. *The Civilian Conservation Corps At Work--Montana and Northern Idaho.* B/W, Silent, 14 min., 35mm, 1934. (A II, #NWDNM(m)-33.472)

2057. _____. *The Civilian Conservation Corps At Work--Mosquito Control.* B/W, Silent, 12 min., 35mm, 1936. (A II, #NWDNM(m)-33.486)

2058. _____. *The Civilian Conservation Corps At Work--Oregon and Washington.* B/W, Silent, 12 min., 35mm, 1934. (A II, #NWDNM(m)-33.471)

2059. _____. *The Civilian Conservation Corps at Work--Utah and Southern Idaho.* B/W, Silent, 16 min., 35mm, 1934. (A II, #NWDNM(m)-33.473)

2060. _____. *The Civilian Conservation Corps Fights Erosion.* B/W, Sound, 11 min., 1935. (A II, #NWDNM(m)-33.488)

2061. _____. *Forest and Men.* B/W, Silent, 10 min., 35mm, 1933. (A II, #NWDNM(m)-33.476)

2062. _____. *Hold That Silt.* B/W, Sound, 17 min., 35mm, 1936. (A II, #NWDNM(m)-33.627)

2063. _____. *Mustard Sowing for Erosion Control.* B/W, Silent, 16 min., 35mm, 1938. (A II, #NWDNM(m)-33.549)

2064. _____. *Safety in Woods Work.* B/W, Sound, 18 min., 35mm, 1938. (A II, #NWDNM(m)-33.552)

2065. _____. *Safety on the Job.* B/W, Sound, 10 min., 35mm, 1938. (A II, #NWDNM(m)-33.550)

2066. _____. *Smoke.* B/W, Silent, 15 min., 35mm, 1931. (A II, #NWDNM(m)-33.397)

2067. _____. *Trout Stream Improvement.* B/W, Sound, 10 min., 35mm, 1937. (A II, #NWDNM(m)-33.520)

2068. _____. *We Can Take It.* B/W, Silent, 31 min., 35mm, 1933. (A II, #NWDNM(m)-33.500)

2069. _____. *Winter Sports in the White Mountain National Forest.* B/W, Silent, 29 min., 35mm, 1933. (A II, #NWDNM(m)-33.528)

2070. United States. Civilian Conservation Corps. *A Day in Virginia Camps.* B/W, Silent, 12 min., 35mm, 1934(?). (A II, #NWDNM(m)-35-3)

2071. _____. *Erosion Control.* B/W, Sound, 10 min., 35mm, 1934(?). (A II, #NWDNM(m)-35.2)

2072. United States. Coast Guard. *Newsreels of the Coast Guard.* B/W, Sound, 10 min., 35mm, 1936(?). (A II, #NWDNM(m)-26.21)

2073. United States. Department of Health, Education and Welfare. *The Army in Overalls: A Defense Report on Film.* B/W, Sound, n/m, 16mm, 1941. (A II, #NWDNM(m)-235.2)

2074. United States. Department of the Interior. *A Forest Playground.* B/W, Silent, 8 min., 35mm, 1936. (A II, #NWDNM(m)-48.34)

2075. _____. *A Nation-Wide System of Parks.* B/W, Sound, 17 min., 35mm, (A II, #NWDNM(m)-48.99)

2076. _____. *A Veteran of Three War's; Old Fort Frederick Maryland Will Live Again.* B/W, Silent, 21 min., 35mm, ca.1939. (A II, #NWDNM(m)-48.106)

2077. _____. *Alabama Highlands.* B/W, Sound, 11 min., 35mm, 1937. (A II, NWDNM(m)-48.68)

2078. _____. *Big Sur, California.* B/W, Silent, 10 min., 35mm, 1937. (A II, #NWDNM(m)-48.59)

2079. _____. *Cradle of the Father of Waters.* B/W, Sound, 10 min., 35mm, 1938. (A II, #NWDNM(m)-48-20)

2080. _____. *Down Mobile Way.* B/W, Sound, 11 min., 35mm, 1935. (A II, #NWDNM(m)48.31)

2081. _____. *Emergency Conservation Work at Mesa Verde.* B/W, Silent, 6 min., 35mm, 1934. (A II, #NWDNM(m)-48.37)

2082. _____. *Great Smoky Mountains National Park.* B/W, Sound, 11 min., 35mm, 1936. (A II, #NWDNM(m)-48.21)

2083. _____. *Hall of Giants.* B/W, Silent, 11 min., 35mm, 1935. (A II, #NWDNM(m)-48.94)

2084. _____. *Heart of the Confederacy.* B/W, Sound, 10 min., 35mm, 1937. (A II, #NWDNM(m)-4833)

2085. _____. *Land of the Giants.* B/W, Sound, 27 min., 35mm, 1935. (A II, #NWDNM(m)-48.32)

2086. _____. *Mount Ranier National Park.* B/W, Silent, 7 min., 35mm, 1936. (A II, #NWDNM(m)-48.81)

2087. _____. *New York, Pennsylvania, Gettysburg, Chickamauga and Moundsville, Alabama State Parks.* B/W, Silent, 11 min., 35mm, 1936. (A II, #NWDNM(m)-48.10)

2088. _____. *Outdoors in the Garden State.* B/W, Sound, 17 min., 35mm, 1937. (A II, #NWDNM(m)-48.36)

2089. _____. *Pilgrim Forests.* B/W, Sound, 19 min., 35mm, 1934. (A II, #NWDNM(m)-48.11)

2090. _____. *Price of Progress.* B/W, Silent, 10 min., 35mm, 1952. (A II, #NWDNM(m)-48.102)

2091. _____. *Rebuilding Indian Country.* B/W, Sound, 36 min., 35mm, 1933. (A II, #NWDNM(m)-48.9)

2092. _____. *Recreation Resources.* B/W, Sound, 15 min., 35mm, 1935. (A II, #NWDNM(m)-48.22)

2093. _____. *Saving the Beauty of Alabama.* B/W, Silent, 6 min., 35mm, 1935. (A II, #NWDNM(m)-48.69)

2094. _____. *Sequoia National Park.* B/W, Silent, 10 min., 35mm, 1938. (A II, #NWDNM(m)-48.87)

2095. _____. *The Shenandoah National Park.* B/W, Silent, 11 min., 35mm, 1937. (A II, #NWDNM(m)-48.104)

2096. _____. *Temples and Peace.* B/W, Sound, 8 min., 35mm, 1937. (A II, #NWDNM(m)-48.3)

2097. _____. *Where Trade Winds Blow; CCC Activities in the Virgin Islands.* B/W, Sound, 21 min., 35mm, 1937. (A II, #NWDNM(m)-48.105)

2098. _____. *Winter Sun and Summer Sea.* B/W, Sound, 21 Min., 35mm, 1938. (A II, #NWDNM(m)-48.35)

2099. United States. Department of the Interior. Bureau of Mines. *Glimpses of Texas*. B/W, Silent, 15 min., 35mm, ca. 1936. (A II, #NWDNM(m)-70.158)

2100. _____. *The Shenandoah National Park*. B/W, Silent, 15 min., 35mm, ca. 1936. (A II, #NWDNM(m)-70.159)

2101. United States. Farm Security Administration. Farmers Home Administration. *The River*. B/W, Sound, 32 min., 35mm, 1937. (A II, #NWDNM(m)-96.1)

2102. United States. Information Agency. *Ocala*. B/W, Sound, 9 min., 35mm, ca. 1945. (A II, #NWDNM(m)-306.123)

2103. United States. International Communications Agency. *Ocala*. B/W, Sound, 9 min., 35mm, ca. 1940-50. (A II, #NWDNM(m)-306.123)

2104. United States. Office of Civilian Defense. *The Enemy Fire*. B/W, Sound, 12 min., 16mm, 1943. (A II, #NWDNM(m)-171.3)

2105. United States. Social Security Administration. *Football Games and News*. B/W, Sound, 10 min., 35mm, 1936. (A II, #NWDNM(m)-47.4)

2106. United States. Works Progress Administration. *A Better Michigan*. B/W, Sound, 8 min., 35mm, 1937. (A II, #NWDNM(m)-69.41)

2107. _____. *A Better Minnesota*. B/W, Sound, 9 min., 35mm, 1937. (A II, #NWDNM(m)-69.44)

2108. _____. *CCC at La Purisima*. B/W, Silent, 11 min., 16mm, ca.1936. (A II, #NWDNM(m)-69.59)

2109. _____. *CCC at Mono*. B/W, Silent, 20 min., 16mm, ca. 1936. (A II, #NWDNM(m)-69.65)

2110. _____. *Flood Special! [West Virginia, 1936]*. B/W, Sound, 7 min., 1936. (A II, #NWDNM(m)-69.8

2111. _____. *March of Time, Special, 1936*. B/W, Sound, 10 min., 35mm, 1936. (A II, #NWDNM(m)-69.9)

2112. _____. *Recovery in Michigan* (Reels 3 & 4). B/W, Silent, 90 min., 35mm, ca. 1937. (A II, #NWDNM(m)-69.51)

2113. _____. *Shock Troops of Disaster*. B/W, Sound, 11 min., 35mm, 1939. (A II, #NWDNM(m)-6949)

2114. Washington State Parks and Recreation Commission. *The Best Time of My Life: A Documentary of the CCC and its Role in the Development of the Washington State Parks*. Color, Sound, 17 min., VHS, 1994.

2115. *Wintertime in the Minnesota Conservation Camps*. B/W, Silent, 20 min., 16mm, 1934-35. [Minnesota: n.d.]

MAPS

(Note of Location: LC#=Library of Congress; NA#=National Archives)

2116. Helms, Douglas. *Civilian Conservation Corps (CCC) Camps, 1933-1942, Technical Supervision by the Soil Conservation Service.* 1:10,000,00. Fort Worth, TX: USDA-SCS, 1983. (LC# 84-693950 MAPS)

2117. Illinois State Planning Commission. *C.C.C. Camps in Illinois.* 1:1,560,000. Chicago: The Commission, 1939.

2118. U.S. Geological Survey. *Civilian Conservation Corps Camps.* 1:2,500,000. [Washington, DC]: U.S.G.S., 1937. (LC# 98-689005 MAPS)

2119. U.S. Soil Conservation Service.

Many maps are dispersed throughout the articles listed. For example, *Reclamation Era* contains maps showing the locations of CCC camps under their control.

PHOTOGRAPHS, STILL PICTURES & SLIDE COLLECTIONS

2120. United States. Bureau of Indian Affairs. *Indian Reservations and Agencies*. 1600 Images, 1927-1940. (A II, #NWDNS-75-RA).

2121. United Stats. Civilian Conservation Corps. *Lantern Slide Set Showing National Youth Administration and CCC Emergency Conservation Work.* 20 Slides, 1938. (A II, #NWDNS-35-LS).

2122. _____. *Photographs of CCC Camps in Connecticut, Massachusetts, and Rhode Island.* 62 Images, July 26, 1933-September 13, 1933. (A II, #NWDNS-35-CA).

2123. _____. *Photographs Related to the Civilian Conservation Corps and Bolling Air Field, Washington, DC.* 373 Images, ca. 1940. (A II, #NWDNS-35-CCC).

2124. United States. Department of Agriculture. Bureau of Agricultural Engineering. Division of Drainage. *Photographs of Civilian Conservation Corps Drainage Projects.* 3000 Images, 1935-1938. (A II, #NWDNS-8-DDR).

2125. United States. Department of Agriculture. Soil Conservation Service. *Photographs of the Civilian Conservation Corps Activities and Projects.* 25 Images, 1936-1940. (A II, #NWDNS-114-B).

2126. United States. Department of the Interior. National Park Service. *Photographs of Display and Exhibit Construction at the Western Museum Laboratory*. 824 Images, 1929-1945. (A II, #NWDNS-79-ML).

2127. _____. *Photographs of Emergency Conservation Work Projects*. 237 Images, 1933-1937. (A II, #NWDNS-79-CCC).

2128. _____. *Photographs Related to the Supervision of Project Training in the National Park Service, and the Civilian Conservation Corps*. 659 Images, 1935-1941. (A II, #NWDNS-79-TR).

2129. United States. Library of Congress. Prints and Photographs Division. Civilian Conservation Corps Collections, Historic American Buildings Survey, Historic American Engineering Record (HABS/HAER) Documentation Collection, and National Register of Historic Places Collection.

2130. United States. Work Projects Administration. *Report of Progress of the Works Program*. 446 Images, 1935-1941. (A II, #NWDNS-69-PS).

CCC PERSONNEL RECORDS

2131. National Personnel Records Center
(Civilian Personnel Records)
111 Winnebago Street
St. Louis, MO 63118-4199

Requests made the Civilian Personnel Records Center must be in writing and proof of relationship to the individual information is being requested on must be provided.

ARCHIVES & REPOSITORIES

2132. Denver Service Center Library and Archives
Bureau of Reclamation Service Center Library
National Park Service
P.O. Box 25007 (D-7925)
Denver, CO 80225
Phone: (303) 445-2072

2133. Franklin D. Roosevelt Library
511 Albany Post Road
Hyde Park, NY 12538
Phone: (914) 229-8114, Fax: (914) 229-0872
E-mail: library@roosevelt.nara.gov

2134. National Park Service Historic Photograph Collection
Harpers Ferry Center,
P.O. Box 50
Harpers Ferry, WV 25425
Phone: (304) 535-6707, Fax: (304) 535-6712

2135. National Park Service History Collection
Harpers Ferry Center
P.O. Box 50
Harpers Ferry, WV 25425
Phone: (304) 535-6262, Fax: (304) 535-6492

2136. Library of Congress: Gifford Pinchot Papers
2137. _____. Harold L. Ickes Papers
2138. _____. Prints & Photographs Division
2139. _____. Monthly Catalog of Publications

2140. National Archives and Records Administration and National Archives Branch Offices:

2141. RG 12: Office of Education
2142. RG 15: Veterans Administration
2143. RG 16: Department of Agriculture
2144. RG 22: U.S. Fish and Wildlife Service
2145. RG 35: Civilian Conservation Corps
2146. RG 48: Office of the Secretary of the Interior
2147. RG 55: Government of the Virgin Islands
2148. RG 69: Works Projects Administration
2149. RG 75: Bureau of Indian Affairs
2150. RG 77: U.S. Army Corps of Engineers
2151. RG 79: National Park Service
2152. RG 95: U.S. Forest Service
2153. RG 114: Soil Conservation Service
2154. RG 115: Bureau of Reclamation
2155. RG 165: War Department
2156. RG 174: Department of Labor
2157. RG 407: Adjutant General's Office

2158. Bancroft Library
Stephen T. Mather Collection
Regional Oral History Collection
University of California, Berkeley
Berkeley, CA 94720-6000
Phone: (415) 642-3773

2159. University of California, Los Angeles
University Research Library
Special Collections department
Horace M. Albright Papers
405 Hilgard Avenue
Los Angeles, CA 90024-1575
Phone: (213) 825-1201

2160. Yale University
Yale University Library
Department of Manuscripts and Archives
John D. Guthrie Papers (1902-1948)
Box 1603 A, Yale station
New Haven, CT 06520

ORAL HISTORY PROJECTS

2161. Bancroft Library
Regional Oral History Collection
University of California
Berkeley, CA 94720-6000
Phone: (415) 642-3773

2162. Brigham Young University Archives
Room 5030
HBLL BYU
Provo, UT 84602
Phone: (801) 378-3514

2163. Columbia University Oral History Research
Box 20, Room 801
Butler Library
New York, NY 10027
Phone: (212) 854-2247

2164. Oral History Collection
Cortland County Historical Society
25 Homer Avenue
Cortland, NY 13045
Phone: (607) 756-6071

2165. Special Collections Department
William Russell Pullen Library
Georgia State University
100 Decatur Street, S.E.
Atlanta, GA 30303-3202
Phone: (404) 651-2185

2166. Department of the Interior
National Park Service
National Park Service History Collection
P.O. Box 50
Harpers Ferry, WV 25425-0050
Phone: (304) 535-6262, Fax: (304) 535-6492

2167. Oral History Program: O.H. 853, O.H. 1311
History Department
Youngstown State University
Youngstown, OH 44555

2168. Southeastern Utah Project O.H. 1108
California State University-Fullerton
800 N State College Blvd.
P.O. Box 4150
Fullerton, CA 92634
Phone: (714) 773-2714

2169. State Historical Society of Wisconsin
Archives Division
816 State Street
Madison, WI 53706
Phone: (608) 264-6400
Web: http://www.shsw.edu/

2170. Utah State Historical Society
Oral History Collection
300 Rio Grande
Salt Lake City, UT 84101-1182
Phone: (801) 533-3500 Fax: (801) 533-3503
Web: http://www.ce.ex.state.ut.us/history/aboutus/aboutus.html
E-mail: ushs@history.state.ut.us

2171. Utah State University Special Collections
Merrill Library
Logan, UT 84322
(801) 750-2637

2172. Wozniak, Toddy. *Civilian Conservation Corps Experiences*. [Fullerton, CA]: Utah State Historical Society and California State University, Fullerton, Oral History Project, 1984.
800 N State College Blvd.
P.O. Box 4150
Fullerton , CA 92634
(714) 773-2714

SPECIAL LIBRARIES

2173. Bureau of Reclamation Service Center Library
Department of the Interior
P.O. Box 25007 (D-7925)
Denver, CO 80225
Phone: (303) 445-2072
(Located in Building 67, Denver Federal Center)

2174. U.S. Department of Agriculture
National Agricultural Library
Public Service Division, Room 111
10301 Baltimore Blvd.
Beltsville, MD 20705
Phone: (301) 344-0257

2175. Department of the Interior
National Park Service
Harpers Ferry Center Library
P.O. Box 50
Harpers Ferry, WV 25425-0050
Phone: (304) 535-6262, Fax: (304) 535-6492

2176. Department of Education/National Education Library
Room 101
555 New Jersey Avenue, NW
Washington, DC 20208
Phone: (202) 219-2289
Fax: (202) 219-1696
Toll Free: (800) 424-1616
Web: http://www.ed.gov/NLE

2177. U.S. Army Military History Institute
22 Ahsburn Drive, Carlisle Barracks
Carlisle, PA 17013-5008
Phone: (717) 245-3611 Fax: (717) 245-3711
Web: http://carlisle-www.army.mil/usamhi/
E-Mail: Archival Collections: MHIAR@awc.carlisle.army.mil
2178. Historical Questions: MHI-HR@awc.carlisle.army.mil
2179. Photos and Artifacts: MHI-SC@awc.carlisle.army.mil
2180. Administrative Info: AWCC-DMH@awc.carlisle.army.mil
2181. Technical Inquiries: STEINKEL@awc.carlisle.army.mil

2182. NACCCA Library
(National Association of CCC Alumni)
P.O. Box 16429
St. Louis, MO 63125-0429
Phone: (314) 487-8666, Fax: (314) 487-9488
Web: http://www.cccalumni.org
E-mail: naccca@aol.com

2183. Franklin D. Roosevelt Library
511 Albany Post Road
Hyde Park, NY 12538
Phone: (914) 229-8114, Fax: (914) 229-0872
E-mail: library@roosevelt.nara.gov

CCC MUSEUMS

2184. Historic Backbone State Park
1347 129th Street
Dundee, IA 52038
Phone: (319) 924-2527
Web: http://www.state.ia.us/parks/backbone.htm

2185. Northeast States Civilian Conservation Corps Museum
166 Chestnut Hill Road (Rt. 190)
Stafford Springs, CT 06076
Phone: Museum (203) 974-1346 Info: (203) 623-3150

2186. Bear Brook State Park CCC Museum
154 Deerfield Road (Off Route 28)
Allenstown, NH 03275
Phone: (603) 485-9874
Web: http://www.nhparks.state.nh.us/parkops/history.html

2187. Civilian Conservation Corps Museum
P.O. Box 13510
San Luis Obispo, CA 93406
Phone: (805) 788-0517
E-mail: cccmuseum@thegrid.net

2188. The Conservation Corps State Museum
c/o Marion Wilbur, Volunteer
5118 San Aquario Drive
San Diego, CA 92109-1510
Phone: (619) 270-5417
E-mail: Budmarion@aol.com

2189. CCC Museum
MCT
P.O. Box 337
Contoocook, NH 03229-0337
Phone: (603) 746-9000

2190. James F. Justin Civilian Conservation Corps Museum
Online Museum Only
Web Address: http://members.aol.com/famjustin/CCChist.html

2191. CCC Museum (Grayling)
North Higgins Lake State Park
P.O. Box 1252
Roscommon, MI 48653
Phone: (517) 821-6374
Web: http://www.sos.state.mi.us/history/museum/museccc.html

2192. Pocahontas State Park CCC Museum
10301 State Park Road
Chesterfield, VA 23838-4713
Phone: (804) 796-4255
Web Address: http://state.va.us/~dcr/parks/pocahont.htm

2193. Watoga State Park
Star Route 1, Box 252
Marlinton, WV 24954
Phone: (800) CALL WVA, 1-800-225-5982

2194. XCCCERS Museum
Joe Toltin, Curator
17833 Jamestown Way
Lutz, FL 33549-6678
Phone: (813) 968-1143

Museums Holding CCC Materials, but not Exclusively CCC

2195. Blue Ridge Parkway Museum
400 BB & T Building
Asheville, NC 28801

2196. Catoctin Mountain Park Museum
c/o Park Ranger
Catoctin Mountain Park
Thurmont, MD 21788
Phone: (301) 663-9399

2197. Chief, Office of Library and Archival Services
Harpers Ferry Center
National Park Service
P.O. Box 50
Harpers Ferry, WV 25425-0050
Phone: (304) 535-6262

2198. West Virginia CCC Museum Association
c/o Dr. Robert Anderson
31 Maple Lake Drive
Bridgeport, WV 26330-9571
Phone: (304) 842-5194

Author Index

Numbers in the index correspond to Bibliography entry numbers in the text.

About the Author

LARRY N. SYPOLT is Senior Project Coordinator at the Institute for the History of Technology and Industrial Archaeology at West Virginia University. He has worked on various history research projects including National Register nominations and historic preservation studies. He serves as an adviser to museums and the Morgantown Historic Landmarks Commission.

www.ingramcontent.com/pod-product-compliance
Ingram Content Group UK Ltd.
Pitfield, Milton Keynes, MK11 3LW, UK
UKHW020808290926
15470UKWH00003B/1028

9 780313 313134